ADVANCES IN ZOOLOGY RESEARCH

ADVANCES IN ZOOLOGY RESEARCH

VOLUME 6

ADVANCES IN ZOOLOGY RESEARCH

Additional books in this series can be found on Nova's website
under the Series tab.

Additional e-books in this series can be found on Nova's website
under the e-book tab.

ADVANCES IN ZOOLOGY RESEARCH

VOLUME 6

OWEN P. JENKINS
EDITOR

nova publishers
New York

For permission to use material from this book please contact us:
Telephone 631-231-7269; Fax 631-231-8175
Web Site: http://www.novapublishers.com

Library of Congress Cataloging-in-Publication Data

ISBN: 978-1-61122-061-2
ISSN: 2162-0962

Published by Nova Science Publishers, Inc. † New York

Contents

PREFACE

This book presents a comprehensive review of various biological and ecological studies in animal science, including their embryology, evolution, classification, habitats and distribution. Topics discussed herein include rodent models of central nervous system tumors; case studies in which butterflies appear in artworks produced by individuals at the very end of life; comparison between biological control with fungi (duddingtonia flagrans and monacrosporium thaumasium) and chemical control (moxidectin) of gastrointestinal nematodiasis in beef cattle; micro-spaces of avian foraging and human presence; a comparative analysis of human perceptions of cougars and other large carnivores in Canada and El Salvador; pattern formation and diversity in butterfly wings; d-amino-acid oxidase in rodents; the raccoon dog (nyctereutes procyonoides) in the community of medium-sized carnivores in Europe; diversity of parasitic nematodes of marine fishes from the gulf of Thailand; and the role of snails in renal epithelial to mesenchymal transition.

Chapter 1 – Rodents are widely utilized for modeling human central nervous system diseases and their use in experimental neuro-oncology continues to increase. The rodent is one of the first animal systems that many new surgical techniques, novel therapeutics and new drug delivery systems are tested on prior to undergoing further pre-clinical analysis. Rodent models are cost-efficient, reproducible, and can be manipulated to mimic the intended disease state. In some cases the inbred status of a rodent can aid in maintaining a controlled experimental environment in which tested variables are manipulated and selective transgenic manipulation can further assist in evaluating therapeutic targets and biological pathways. Alternatively, immuno-compromised animals can be used to host tissue from other species, such as human tumor explants, to more effectively test novel drugs or treatment combinations for intended future clinical use. In this chapter the authors identify some of the most utilized rodent models for primary brain and spinal cord tumors as well as some models that have been implemented to study metastatic disease to the brain and the spinal column.

Chapter 2 – As a professor of the humanities at Rice University, and as an Artist In Residence in the Department of Palliative Care and Rehabilitation Medicine at the University of Texas M. D. Anderson Cancer Center, the authors has had the opportunity to witness the many dynamic qualities of butterflies. These remarkable beings appear at once as figures of strength and delicacy, freedom and interconnectedness, endurance and fragility. Inhabiting the terrestrial and the aerial domains simultaneously, butterflies are closely interwoven with their natural and social environments, just as they remain distinctive presences within them. In this commentary, the author presents several vivid case studies in which butterflies appear in artworks produced by individuals at the very end of life. These moving words and images exemplify the ways in which butterflies embody key themes within aesthetics, spirituality, and the medical humanities. Throughout the stories butterflies appear as figures of transience and endurance that cross multiple boundaries within and beyond the end of life. These materials thus offer an innovative perspective that complements the views of the scientific community, while contributing to a larger case for the environmental importance and cultural value of the interconnected lives of human beings and butterflies.

Chapter 3 – Sodium alginate pellets of two different nematode predatory fungi *Duddingtonia flagrans* and *Monacrosporium thaumasium* (biological control) and moxidectin (chemical control) were evaluated in the control of gastrointestinal nematodiasis in bulls. Four groups of ten 6-month-old male Nellore calves were kept in paddocks of *Brachiaria decumbens* for 12 months. Each animal of group A received 1 g/10 kg body weight (b.w.) of pellets of *D. flagrans* (0.2 g fungus/10 kg b.w.) twice per week, for 12 months. In group B, each animal received 1 g/10 kg b. w. of *M. thaumasium* pellets, twice per week, for 12 months. In group C, each animal received 1 ml/50 kg b.w. of the anthelmintic moxidectin (1%) in May, July and September 2010. Animals of the control group (group D) received no treatment. In the analysis of fecal egg count (eggs per g feces; EPG), groups A and B had a 56.67% and 47.8% smaller count, respectively, than that of the control group ($p<0.05$). Moxidectin reduced EPG by 93.8% in the first 6 months of the study (period of application) and by 12.4% in the last 6 months, indicating reinfection by nematodes. Biological control maintained a low EPG in the long term while chemical control resulted in a short-term effect. The percentage reduction of infective third-stage larvae (L_3) in relation to the control group at distances of up to 20 and 20–40 cm from the fecal pat was 64.5% and 73%, respectively, for group A; group B showed percentage reductions of 47.3% and 58% at the same distances. *Duddingtonia flagrans* showed a larger percentage reduction of L_3 in the pasture compared with *M. thaumasium*. Moxidectin showed no reduction of larvae in the pasture. Treatment of bulls with nematophagous fungus and moxidectin integrated in pellets can be used as an alternative treatment of bovine gastrointestinal nematodiasis; however, biological control was more efficient than chemical control over the long term in the environmental conditions of the present study.

Chapter 4 – Traditional zoogeographical theory focuses on global regions, with less attention to small spaces shared by people and animals. Two relevant, possibly opposite, research strands, the statistical analysis of alert/flight distances and the subjective 'actancy' strand within animal geography are rarely examined as complementary to a new look at micro-spaces of human/avian relations within the larger theoretical framework of zoogeography. Alert distances (the distance between a bird and an approaching human when the bird shows awareness), flight distances (the distance, in such a scenario when the bird takes flight) and alert periods (the period between first alertness and flight) vary according to cover structure, elevation and proximity, human and road proximity. This chapter argues that while these parameters give a generalized contribution to the sharing of micro spaces, a strand of animal geography that emphasizes the unpredictable, subjective, individualist behavior of animals (birds in this study) offers a complementary perspective to the more classical statistical prediction of alert/flight distance based analyses. Illustrative examples are four papers published by this author; two on the alert distances of birds in the green spaces of Stirling, Scotland and Peterborough, Canada; two papers on the utility of a strand of animal geography (emphasizing avian individualism) in feeding spaces of Glasgow, Scotland and

Peterborough, Canada; and one paper on the landscape ecology of birds in Ottawa, Canada. It is concluded that for effective environmental applications, the study of alert and flight distances must acknowledge bird (intra- and inter- species) adaptation and habituation to human presence; the lack of such acknowledgment may reduce the validity of distance analysis in avian studies, with implications for conservation and landscape planning. This represents an important contribution to the sub-discipline of zoogeography and to its parent disciplines of zoology and geography.

Chapter 5 – The relationship between people and the cougar *Puma concolor* and other larger predators in Canada (brown bears *Ursus arctos*, black bears *Ursus americanus*) and Latin America (jaguars *Panthera onca*) is examined in a comparative mode for cases in Kamloops and Vancouver Island, Canada and El Salvador. Literature derived hypotheses mooting greater sympathy for carnivores from women and young people, greater fear from women and older people, and greater fear of the larger brown bears and jaguars over cougars are tested. The results show that the hypotheses are both supported and refuted. The size of the predator is as important as common ideas about its behaviour, and troublesome invasive behaviour (mostly from bears) is at least as important as predatory behaviour (largely from cougars and jaguars). However, consistent with hypotheses, women and older people were in many instances more afraid of cougars and other carnivores, and older people were generally less tolerant of carnivores. Differences included opinions on positive impacts on quality of life and useful functions of cougars (favoured by men in El Salvador, both men and women in Vancouver Island, and women in Kamloops), and danger to adults, children and other animals (favoured by women more than men in El Salvador for all groups, women more than men in Vancouver Island for children and animals, not adults, and women more than men in Kamloops for adults and children, not animals). For age, older people in El Salvador were generally less tolerant than younger people of large carnivores, similar to Vancouver Island, but not Kamloops, where there were no age-related opinions of the danger from carnivores towards people. The cougar's elusive behaviour limited comparisons with bears and jaguars. Awareness of such issues may assist wildlife conservation.

Chapter 6 – Butterfly wings are covered with thousands of monochromatic scale cells, which form regularly arranged patterns, that is, parallel rows of scale cells in the proximal-distal direction of the wing. Color pattern in butterfly wings is a kind of mosaic pattern of dozens to hundreds of finely-tiled monochromatic scale cells. The wing consists of two cell layers, a surface layer and a back layer. The two layers are completely separated by the middle tissues, and developmental phenomena in each layer occur independently of the other. Color pattern formation occurs in a single layer of the wing disc and it is essentially two-dimensional in space without cell movement. Therefore, the problem of color pattern formation could be solved by analyzing the underlying mechanism of how colored scales are settled at specific positions in a two-dimensional wing surface. In this article, The author reviews the progress of experiment-based mathematical modeling for color pattern formation and diversification in butterfly wings. First, the author describes fundamental features of the wing color pattern, and then goes into discussions on guiding principles or governing rules to derive from the so-called ground plan to diverse wing color patterns. In order to make the problem clear, the author takes up some specific butterfly species as model systems, of which color pattern and its diversity have been well investigated with both experiment and mathematical analysis. Finally, the author would like to stress the importance of integration of different aspects such as gene and environment for understanding of the diversity in wing color patterns.

Chapter 7 – D-Amino-acid oxidase (DAO) catalyzes oxidative deamination of unnatural D-amino acids, producing the corresponding 2-oxo acids, hydrogen peroxide and ammonia. DAO is present in a wide range of organisms from yeasts to humans. In mammals, it is present in the brain, liver and kidney. The number of amino acid residues consisting of DAO is 347 in the human, pig, and rabbit. The guinea pig DAO has 347 amino acid residues whereas DAOs of the rat and hamster consist of 346 amino acid residues. Furthermore, mouse DAO has 345 amino acid residues. Thus, the numbers of amino acid residues are different depending on the species of rodents. Accordingly, the authors searched mammalian DAOs, especially rodent DAOs, in databases and examined their amino acid sequences. The authors found that DAOs of monkey, cattle, squirrel, and degu have 347 amino acid residues and jerboa DAO has 346 residues whereas the prairie vole DAO has 345 residues. Comparison of amino acid sequences has indicated that the number of the amino acid residues and the homology of amino acid sequences are clearly parallel to the taxonomic positions of the rodents.

Chapter 8 – Raccoon dog (*Nyctereutes procyonoides*) is an alien carnivore in Europe. It is today widespread and common in many parts of the continent. Here the authors review its adaptations to various environmental and climatic conditions, impact on the native fauna and management of the populations. The raccoon dog is very adaptable due to its versatile diet, ability to gather large fat reserves in autumn and to hibernate in areas where winters are harsh. It also has higher than expected reproductive potential, which compensates the high mortality rate of juveniles. Raccoon dogs are monogamous and the male participates in pup rearing. Since no helpers have been observed, juveniles usually disperse during their first autumn and some of them wander far. In a colonizing population also adults have a tendency to wander long distances. Due to massive and dispersed introductions, raccoon dogs are characterized by high genetic variation. Raccoon dogs fitted well into the guild of medium-sized carnivores and share their home ranges with badgers (*Meles meles*) and foxes (*Vulpes vulpes*), and especially raccoon dogs and badgers partly use the same habitats, such as deciduous forests and fields. Raccoon dogs often settle in active badger setts to avoid predation and unfavourable weather conditions. Raccoon dogs may partly overlap their diet with that of native carnivores. Raccoon dogs can spread diseases and parasites, such as rabies and sarcoptic mange, to other carnivores due to their frequent opportunities of contact. The raccoon dog is a true omnivore and eats anything it can find and catch. Although it sometimes preys on birds, there is no firm evidence that it affects bird populations. The occurrence of birds in its diet shows great spatial and seasonal variation but in most areas is fairly low. Raccoon dogs may have, however, an impact on amphibian populations, especially on islands. Raccoon dogs are actively hunted in many countries. In Finland, the annual hunting bag is around 160,000 individuals, i.e., about 50% of the autumn population. In spite of many campaigns to reduce the population, the annual hunting bag has steadily increased during the past two decades in Finland. The hunting bag has also increased exponentially in Germany since the early 1990s. Due to their adaptability, large distribution and secretive life, raccoon dogs are very difficult to eradicate after they have established in an area.

Chapter 9 – Eleven genera of parasitic nematodes of marine fishes from the Gulf of Thailand were reported. Five genera of larval Ascaridoid nematodes (*Anisakis, Contracaecum, Terranova, Raphidascaris* and *Porrocaecum*) were found in the mesenteries, visceral organs and intestines of many species of marine fishes. *Anisakis* was found in 4 species, Type I Anisakis larvae were found in 22 species, larvae of the genus *Contracaecum* were found in 2 species, and type A *Contracaecum* larvae were found in 12 species. Type B *Contracaecum* larvae were found in 5 species. Type B *Terranova* larvae were found in 13 species. Larvae of the genus *Raphidascaris* were found in 8 species. Larvae of the genus *Porrocaecum* were found in 2 species. Six genera of mature parasitic nematodes, namely *Capillaria, Cucullanus, Dichelyne* (*Cucullanellus*), *Camallanus, Procamallanus* (subgenus *Spirocamallanus*) and *Philometra*, were also reported in the Gulf of Thailand. *Capillaria* sp. was found in one fish species. *Cucullanus* spp. were found in 7 species, whereas *Dichelyne* (*Cucullanellus*) sp. was found in only one species. Genus *Camallanus* was found in 6 species. *Procamallanus* (subgenus *Spirocamallanus*) was found in 13 species. *Philometra* was found in 11 species. Details regarding parasitic nematode distribution in the Gulf of Thailand were reported, and this distribution was compared with that of another location in this chapter.

Chapter 10 – Epithelial-mesenchymal transition (EMT) is a highly controlled process involved in organogenesis, that is increasingly recognized as an integral part of tissue repair, organ fibrosis and cancer progression. *In vitro* data have shown that EMT is regulated by a complex signalling network. Recent *in vivo* data, in particular using tracing technologies, have conclusively demonstrated EMT in fibrotic kidney disease, and confirmed that EMT contributes to the pool of interstitial matrix-producing fibroblasts. In this review, the authors will summarize the known signalling pathways regulating EMT in renal proximal tubule cells following divergent insults, including exposure to high glucose, transforming growth factor β (TGF-β), angiotensin II and hypoxic conditions. This authors will also describe the signalling pathways/mediators differentially involved in the regulation of EMT in the fibrotic kidney, including Notch, Smad, integrin-linked kinase, Wnt/β-catenin and Erk1/2 and their convergent effects on Snail. The potential role of Snail as a common 'master switch' integrating various signal inputs in controlling EMT, its role in different models of renal pathology and potential therapeutic implications will be discussed.

In: Advances in Zoology Research. Volume 6
Editor: Owen P. Jenkins
ISBN: 978-1-61122-061-2

Chapter 1

RODENT MODELS OF CENTRAL NERVOUS SYSTEM TUMORS

***Antonella Mangraviti*[1], *Gustavo Pradilla*[2], *Noam Schildhaus*[1] *and Betty M. Tyler*[1,*]**

[1]Department of Neurosurgery, Johns Hopkins University School of Medicine, Baltimore, Maryland, US
[2]Department of Neurosurgery, Emory University, Atlanta, Georgia, US

ABSTRACT

Rodents are widely utilized for modeling human central nervous system diseases and their use in experimental neuro-oncology continues to increase. The rodent is one of the first animal systems that many new surgical techniques, novel therapeutics and new drug delivery systems are tested on prior to undergoing further pre-clinical analysis. Rodent models are cost-efficient, reproducible, and can be manipulated to mimic the intended disease state. In some cases the inbred status of a rodent can aid in maintaining a controlled experimental environment in which tested variables are manipulated and selective transgenic manipulation can further assist in evaluating therapeutic targets and biological pathways. Alternatively, immuno-compromised animals can be used to host tissue from other species, such as human tumor explants, to more effectively test novel drugs or treatment combinations for intended future clinical use. In this chapter we identify some of the most utilized rodent models for primary brain and spinal cord tumors as well as some models that have been implemented to study metastatic disease to the brain and the spinal column.

* Corresponding Author: Betty Tyler. Department of Neurosurgery, Johns Hopkins University School of Medicine, 1550 Orleans Street, CRB2- 2M45, Baltimore, MD 21231. Tel: 410-502-8197, Fax: 410-614-0478, E-mail: btyler@jhmi.edu.

Brain Tumor Models

Rodent Models

Rodent glioma models are well established and widely utilized in brain tumor research. Rodent models have contributed significantly to the understanding of in vitro and in vivo molecular characteristics and biological behavior of brain tumors, as well as the in vivo effects of experimental treatment options. [1, 2] Certain characteristics are necessary for suitable and useful glioma modeling, including aggressive tumor growth, reliability of tumor growth, infiltrative characteristics, heterogeneous morphology and complete lethality. The tumor model should also have reproducible histo-biology, demonstrate non-encapsulation, and have angiogenic properties. [2, 3]

Although no rodent model possesses all of these qualities, models have been useful investigational tools for decades, and have significantly contributed to the understanding of malignant glioma. We discuss some of the most utilized central nervous system tumor models, including animal tumor models generated in rodents and human tumor xenografts implanted in immuno-compromised rodents.

Syngeneic Rat Models

C6 Glioma

The C6 glioma was developed at the Massachusetts General Hospital, in Dr. William Sweet's laboratory by Benda and Schmidek et al. by repetitive transplacental injection of methylnitrosourea over a period of 8 months to outbred Wistar-Furth rats. [4] Tumor tissue was removed, expanded into tissue culture, and cloned. Following histopathological analysis, the tumor was characterized as an astrocytoma, positive for glial markers such as GFAP and S-100, and it was accessioned by the American Type Culture Collection, Manassas, VA. [2, 3] The C6 glioma cell line is the most highly cited rodent brain tumor model on the National Center for Biotechnology Information's PubMed database for biomedical literature. Implantation models of C6 glioma cells more closely resemble the infiltrative nature of human glioblastomas in the Wistar rat, than any other rodent strain. [3]

Like naturally occurring glioblastomas, C6 glioma implantation results in the formation of a tumor, surrounded by reactive gliosis, with tumor cells showing a high mitotic index, nuclear polymorphism, tumor necrosis, intratumoral hemorrhage, and invasion of brain parenchyma. [2-4] Although the C6 glioma is popular for a variety of investigational studies, particularly those related to understanding tumor biology, this tumor model has limited potential for use in survival studies because its development in an outbred Wistar rat leaves no syngeneic host in which this cell line can be propagated. [2] Due to this, the C6 tumor can be immunogenic in allogenic hosts, limiting its utility.

9L Gliosarcoma

The 9L gliosarcoma, like the C6 glioma, was produced by Benda and Schmidek in Dr. William Sweet's laboratory at the Massachusetts General Hospital, by intravenous injection of methylnitrosourea (5 mg/kg) for 26 weeks in CD Fischer rats. [5, 6]

Following accession and cloning by Barker et al. at the Brain Tumor Research Center, University of California, San Francisco, it was named the "9L Gliosarcoma" due to its sarcomatous appearance. [7] Unlike other glioma models, the 9L gliosarcoma is not as representative of naturally occurring human gliomas, due to its decreased ability to invade brain parenchyma and its inability to induce adequate angiogenesis. [3] In addition, several studies have demonstrated that the 9L tumor is immunogenic. Therefore, syngeneic rats can be immunized against both subcutaneous and intracerebral challenges using irradiated tumor cells. [2, 3] Due to amplification of tumor-associated immune response, this model poses significant limitations for the study of certain treatment modalities, including gene therapy, where tumor cells are selectively killed and immune effector cells are spared. [2, 3] However, the 9L gliosarcoma is a particularly useful brain tumor model in experimental neuro-oncology and preclinical studies because it can be propagated both in vitro and in vivo, thereby increasing its ease and variety of use.

F98 Glioma

The F98 glioma was developed at Ohio State University by a single dose intravenous administration of ethylnitrosourea (50 mg/kg) to a pregnant CD Fischer rat at 20 days of gestation. [8] Histopathological analysis has designated the F98 glioma as an anaplastic or undifferentiated glioma, with an infiltrative growth pattern, with no metastatic component. [9] The F98 glioma is highly aggressive and lethal, and unlike the 9L gliosarcoma, F98 tumors are weakly immunogenic. [10] In addition, similar to human malignant glioma, this tumor has not been cured with conventional soluble nitrosoureas or irradiation. The F98 glioma, therefore, is a suitable model for the study of chemo-resistance to the most widely used chemotherapeutic agents in glioma therapy: alkylating agents. The F98 rat glioma, unlike other cell lines, has a moderate-to-high methylguanine-DNA methyltransferase (MGMT, AGT) activity that has been shown to be directly related to the capacity of tumor cells to remove the O(6)-alkylguanine DNA adducts formed by the most potent cytotoxic DNA-damaging chemotherapeutic agents, such as carmustine (BCNU), temozolomide, streptozotocin, and dacarbazine. [11-14]

The F98 glioma can be considered an ideal model for unresponsive and recurrent gliomas.

RG2 Glioma

The RG2 tumor was developed through the intravenous administration of ethylnitrosourea in a pregnant Fischer rat at 20 weeks gestation. [2]

The tumor was subsequently cloned by Wechsler in Germany and one of the clones was named the "RG2" (rat glioma 2) glioma. [15] The same clone was referred to as the "D74-RG2" or "D74" in Koestner's laboratory at Ohio State University. [2] The RG2 glioma is not immunogenic in syngeneic hosts and it displays an invasive pattern of growth, making it a good candidate for glioblastoma research. [2]

CNS-1 Glioma

The CNS-1 glioma was induced in an inbred Lewis rat that had received weekly intravenous methylnitrosourea injections for six months. [16] The intracranial tumor that subsequently developed was isolated and placed into culture.

Intracranial implantation of the CNS-1 glioma rendered a growth pattern showing single cell parenchymal and leptomeningeal infiltration, along with perivascular and periventricular spread with expansion into the choroid plexus stroma. [2] The CNS-1 glioma was shown to be positive for a number of molecular markers, including GFAP and S-100, which are not present in the 9L gliosarcoma cell line. [16]

Also, in comparison to the 9L gliosarcoma, the CNS-1 glioma has a more infiltrative growth pattern, which more closely resembles human gliomas. [2]

N29 Glioma

The N29 was begun with the administration of ethyl-N-nitrosourea in Fischer rats that were 17-18 days pregnant. After studying the rodents for over 200 days it was found that 80-90% of the offspring had tumors in the central or peripheral nervous system. This tumor was then successfully propagated in vitro and in vivo. [17]

BT4C Glioma

The BT4C rodent glioma is a highly vascularized tumor with infiltrative characteristics. The BT4C cells were implanted into BDIX male rats using 15-LO-1 gene therapy combined with HSV-tk gene therapy and showed that the combination neither helped the tumor survive nor inhibited its growth. The cell line reacted with a migratory effect instead. It is widely used because of its infiltrative and aggressive characteristics. [18]

N32-N29 Glioma

The rat glioma cell lines N29 and N32, syngeneic with the Fischer 344 rat were originally induced by transplacental injection of ethyl-N-nitrosourea into pregnant rats whose offspring then subsequently developed malignant brain tumors. [19] It has been shown that N29 and N32 rat tumor cells grow as tumor spheres, express GBM-associated stem/progenitor cell markers, are clonogenic *in vitro* and highly tumorigenic *in vivo.* The N29 and N32 glioma models are considered homogenous populations of clonogenic and tumorigenic cells. These models are believed to be constituted of a pure population of CD133+ and Nestin+ brain tumor initiating cells. [20]

T9 Glioma

The original stock of T9 glioma cells was obtained from Dr. William Sweet's laboratory at the Massachusetts General Hospital by Denlinger, and Koestner. [2] Although not fully appreciated, the T9 glioma was at one time, and still may be the same as the 9L gliosarcoma. [2] Similar to the immunogenicity of the 9L gliosarcoma, the T9 glioma was also found to be highly immunogenic.

It was found that rats immunized with irradiated T9 cells or T9 cells mixed with Cryptosporidium parvum rejected subsequent subcutaneous implants of T9 glioma cells. [21] In order to immunize against intracranial tumors, rats initially had to reject intradermal T9 cells. The T9 cell line has been used for many studies, including the evaluation of anti-angiogenic and chemotherapeutic agents, immunotherapy, and gene therapy with interferon. [22, 23]

RT2 Glioma

The RT2 glioma model was derived by intracranially inoculating neonatal Fischer 344 rats with purified avian sarcoma virus. [24] This model has been used to study the effects of chemotherapy, blood brain barrier disruption, tumor permeability, gene therapy and radiosensitization. [25] The RT2 tumor appears to be immunogenic, as evidenced by its ability to evoke a CD8 + T cell-mediated anti-tumor immune response, a fact which must be considered if used for immunotherapy studies. RT2 glioma expressing green fluorescence protein (GFP) has been used for quantitative assessment of intracranial tumor invasion. [26] This model has also been used to evaluate the therapeutic efficacy of oncolytic adenoviruses.

Generation of Intracranial Rat Tumor Models

As outlined above syngeneic rodent glioma models have been primarily developed by direct chemical mutagenesis and maintained in cell culture or subcutaneous passages. [3] Stereotactic implantation of glioma has improved the tumor growth yield, made rodent mortality more uniform and made the more quantitatively standardized cell suspensions could be transplanted, and rodent mortality rate and tumor growth yield improved. [3] Currently, with the development of new implantation techniques, intracerebral tumor growth has reached yield rates close to 90–100%, and animal mortality, as well as the rate of extracranial and spinal metastasis, have significantly decreased. [3] This is due to various advances in transplantation techniques, including augmentation of cell suspensions with 0.5% to 1.0% agarose to prevent backflow of tumor cell into the injection tract, more concentrated cell suspensions, improved injection needles, and flushing of the operative field, as well as better understanding of stereotactic localization. [3] Continuing efforts to improve cell and tissue transplantation will allow for even more precise, reproducible, and easily accessible modeling of intracerebral tumors for experimental studies.

Comparison of the Various Rat Models

Each of the previously described rodent glioma models possesses certain advantages and limitations. The optimal model selected for a specific study depends upon the nature and the purpose of the experiment, with some models being better suited for particular types of investigational studies rather than others. The 9L gliosarcoma model, one of the most popular rodent glioma models, has been widely used to study tumor biology as well as therapeutic efficacy of various treatment modalities. [2]

Many studies use the 9L model to assess glioma characteristics, including cellular properties and metabolism, [27, 28] tumor imaging, [29] drug pharmacokinetics and drug resistance mechanisms, [30] and anti-angiogenic properties of various agents, [31] among others. In addition it has been used to assess the efficacy of different treatment options, including chemotherapy, [32, 33] radiotherapy [34], immunotherapy, [16, 35] and combination therapy. [36] However, due to its immunogenic nature, the 9L tumor is a suboptimal model for the study of treatments such as gene therapy and boron-neutron capture, where cure rates are believed to be due to host immune response amplification. [2] Less immunogenic tumors such as the F98 glioma have been more useful in the study of such treatment options to study the molecular genetic alterations in GBMs, effects of infusion rates on intracranial drug distribution, and for suicide gene therapy with Herpes simplex virus-1 thymidine kinase (HSV-TK). [37, 38]

In addition, F98 tumors have also been utilized to evaluate tumor metabolism and cellular physiology, [28] tumor imaging, [37] tumors not responsive at alkylating agents, [12] and the efficacy of biodegradable polymers for sustained release of cytotoxic drugs. [38, 39] The C6 glioma, derived from outbred rats, is allogenic in all inbred rat strains, limiting its utility as an in vivo tumor model. However, it has been particularly useful in the study of tumor biology, including experiments investigating tumor growth patterns, [40] cellular physiology, [41] and angiogenesis. [42] The RG2 glioma has been used in various studies, including experiments exploring vascular permeability and blood flow, [43, 44] and blood-brain-barrier characteristics. [15] Apart from studies to evaluate therapeutic efficacy, the BT4C glioma model also has been used to study the molecular and biological changes induced by chemotherapy, [45] radiation therapy, [46] and suicide gene therapy. [47] It has been used for a number of studies, including the evaluation of antiangiogenic, [22] and chemotherapeutic agents, [23] immunotherapy, and gene therapy with interferon-beta. [48]

Syngeneic Mouse Model

Murine syngeneic intracranial glioma models are well established and are useful tools for exploring brain tumor biology, molecular causative genetic alterations in tumorigenesis, the therapeutic role of various anti-neoplastic agents, and identification of new targets for treatment. [49] The modeling of murine gliomas occurs through two main approaches: development of tumors by direct exposure to chemical or viral mutagens, or by genetically engineered mutations in germline or somatic cells. [49] Both methods have strengths and drawbacks, with both having a wide range of applications in experimental neuro-oncology. [49]

GL261

The GL261 mouse model is well established and was developed in a B6 mouse by intracranial exposure to methylcholanthrene. [50] The cell line has been available since 1970 through the National Cancer Institute Microbiological Associates (Bethesda, MD). Implantation of GL261 cells into C57BL/6 mice demonstrates highly reproducible patterns of growth and disease progression, with invasion and infiltration of brain parenchyma. [50] The GL261 mouse model shares many histopathological markers with human malignant gliomas and expresses markers of invasiveness, such as CXCR4. [51] The tumor expresses areas of angiogenesis and hypoxia, which correlates with the expression of VEGF and HIF-1α. [52] A key feature of this model is the diffusely infiltrating and invasive characteristics of GL261 cells into normal brain. [52]

Relevant Studies

Tumor models based on orthotopically transplanted tumor cells (tumor placed into the same organ of origin), such as the GL261 glioma, allow for a variety of investigational studies to be conducted in a highly reproducible, predictable manner. The GL261 glioma has been used in several studies including those exploring tumor biology and gene therapy, [53] as well as immunotherapy. [54] This model may mimic most closely the growth and immune response of human GBM.

Indeed, it has made advances in GBM experimental immunotherapy possible on account of it being immune competent and because of the large number of antibodies and immune markers available for mice. [55] GL261 is a strong model for studying therapeutic options for malignant glioma, particularly immune-based ones. [52]

Genetically Engineered Mouse Glioma Models (GEMMs)

Specific genetic alterations have been associated to different subtypes of malignant glioma. Most malignant gliomas appear without previous evidence of a less malignant lesion; these tumors are called primary glioblastomas. Secondary glioblastomas arise from a pre-existing less malignant lesion (i.e., a low grade glioma). [56] Alterations such as amplification of epidermal growth factor receptor (EGFR), [57] mutations or deletions of *NF1* gene, [58] INK4a-ARF deletions, [59] and inactivating PTEN mutations [60] have been associated with primary gliomas. Mutations of p53, CDK4 amplification, *RB* gene loss, [59] and metabolic enzymes like isocitrate dehydrogenase 1 and 2 have been associated mainly with secondary GBMs. [56] Finally, mutations of P16, RB, p14, Cyclin D, and PI3K are shared between both subtypes. [62] Also, platelet derived growth factor receptor (PDGFR) amplification and over-expression has been found in a small subset of GBMs. [62]

Generation of Genetic Models of Glioma

Even though xenograft models of gliomas have been the main tool for in vivo studies of glioma biology, these models rely on the use of established cell lines and primary cell lines cultured from tissues of patients with glioblastoma. [63] With the identification of important mutations that determine gliomagenesis, transgenic mouse models that harbor these mutations have been created. Different techniques have been used to generate these mouse models. Introduction of germ line mutations into embryonic stem cells, gene transfer using engineered viral vectors, homologous recombination for specific gene deletion have been used and are reviewed elsewhere. [64]

Germ line mutation models allow for the study of gain of function or loss of function of certain genes important for the development of gliomas and they can be cell-type specific, since the expression of the transgenes is driven by a determined promoter, such as GFAP promoter for expression in astrocytes or nestin promoter for neural stem cells. [65] Ablation of the expression of certain genes can also be studied using these techniques and also in specific cell populations by the generation of knockout mice, introduction of germ-line heterozygosity, or induction of haploinsuffiiciency. [66, 67] Gene targeting in specific cell populations can also be done through the use of a topoisomerase present in P1 bacteriophage that is able to catalyze recombination of DNA flanked by loxP sites, called Cre recombinase or Cre. [68]

In this type of genetic model, a gene of interest or an important segment of its coding sequence is flanked by loxP sites. After this procedure generation of a conditional knockout can be done by either injecting a virus that expresses Cre into a desired area of the brain or by crossing mouse lines that express Cre under a cell specific promotor with mouse lines that have loxP sites flanking the sequence of interest in their genome. These models can be used to activate the expression of the target gene, through removal of an artificial stop codon or an interfering sequence; also, they are useful for the inactivation of the expression of the target gene through removal of an essential sequence of the gene of interest. Examples of this system include targeting of NF1, PTEN, and p53. [69, 70]

Another method to introduce mutations into targeted cell populations using somatic gene transfer is a system that uses specific promoter driven expression of the receptor for the replication-competent avian leukosis retrovirus splice acceptor viral vectors (RCAS). This system has been used to overexpress a constitutively active EGFR mutant and generate lesions that share many characteristics with gliomas in mice with disrupted INK4a-ARF to induce tumor formation and also to overexpress basic Fibroblast Growth Factor (bFGF) to study its effects on glial cell biology. [59, 71]

Another study used CMV (cytomegalovirus) vectors as promoter; the injection of CMV-Cre into p53;Ink4a/Arf;K-Ras^{v12} resulted in high grade gliomas formation. [72] These mouse models are especially important to study molecular signaling pathways implicated in the pathogenesis of gliomas and to allow studies that require induction of loss of function, gain of function, and deletions of specific genes.

Conclusion

GEMMs, though not uniformly suitable for preclinical testing, provide an accurate recapitulation of the casual genetic events and subsequent molecular evolution of brain tumors. Furthermore, some of them produce tumor-stroma interactions similar to those found in native tumors. They also appear to contain the cellular subpopulations, such as cancer stem cells (CSCs), believed to be essential for the development, maintenance, and drug resistance of brain cancer.

For these reasons, GEMMs offer a powerful toolbox for addressing many of the pressing issues in modern cancer biology, such as the molecular mechanisms and cellular origins of neoplastic processes and the importance of the tumor microenvironment. The importance of genetic models of glioma lies in that known mutations that occur in human gliomas are introduced as the tumorigenic stimulus. This methodology enables the researcher to study the progression of genetic and biological changes that occur as a consequence of these mutations.

However, this implies that the genetic alterations discussed above are the initiators of tumorigenesis. The study of the interaction between tumor cells and the immune system is enhanced using these models, due to the fact that tumors formed express the same major histocompatibility molecules as other cells in the body, eliminating an important confounding factor in the evaluation of immunologically-based therapies.

Disadvantages of stereotactic intracranial injection of glioma cell lines include the genetic alterations that cells acquire with prolonged periods of in vitro culturing even of initially non-tumorigenic NSCs. [73]

Hence, prolonged culture of glioma cells can translate into aberrant genetic alterations that do not represent those seen in the development of gliomas, a problem avoided by the use of genetically engineered models of glioma.

The different genetic manipulations that produce glial tumors in these mice empower the researcher to test therapies in the context of mutations seen in different types of gliomas, i.e., mutations seen in primary glioblastomas versus mutations seen in secondary glioblastomas.

Such experiments will potentially yield important results towards the optimization of therapies for different tumors according to their main genetic alterations and biology. By the use of animal models that represent and recapitulate more accurately the human disease more meaningful information will be generated and applied in the design of improved therapies for GBM.

Athymic Rodents-Human (Xenograft) Models

Experimental brain tumor models need to closely resemble naturally occurring human gliomas, including properties of parenchymal invasion, perivascular migration, and angiogenesis. [74] However, no rodent and murine syngeneic glioma model completely recapitulates these characteristics, therefore, the use of human malignant glioma cells in both in vitro and in vivo studies have been implemented. They are another tool that can determine growth progression, tumor biology, and response to therapeutic options.

Human Cell Lines

Multiple human glioblastoma cell lines are currently available for investigational studies. There are considerable biological and histological differences among the various lines available, mirroring the heterogeneous nature of human brain tumors. [3]

U251 Glioma

The U251 glioma is a commercially available cell line that was derived from the tumor of a 75 year old male with a human malignant glioblastoma. [75]

Adherent growth characteristics of this human glioblastoma cell line have previously been shown to be inhibited by HET0016. [75] U251 recapitulates the salient histological and immune-histochemical and radiological features of human malignant glioma. In addition, a number of genetic alterations show similarities to human malignant glioma, including alterations in key tumor suppressors and oncogenic pathways.

Morover it has been shown that this line contains a $CD133^+$ subpopulation of cells able to form neurosphere aggregates with self-propagating potential. [76] The U251 cell line has been used both in subcutaneous and intracranial mouse models, but it has been reported that there is a gene expression profile differences between the two tumor models, noting the importance of different *in vivo* growth conditions and the role of the microenvironment in these cells. [52]

U373 Glioma

The U373 cell line is commercially available and is described as an anaplastic astrocytoma derived from a 61-year old male. [77] First, the U251 cell line was distributed as U373 MG until short tandem repeat (STR)-PCR profiling confirmed identity with U251. Then a new deposit of U373 MG, known as U373 MG (Uppsala), was made available with its own catalog number. Orthotopic xenograft models yield a glioma that is diffusely infiltrative into the brain parenchyma, with high mitoses, pseudopallisades, and large necrotic centers. [78] The response of these tumors to conventional nitrosourea treatment results in only a marginal increase in survival, similar to naturally occurring human gliomas. [78]

U87 Glioma

The U87 glioma cell line is similar to other human glioblastoma cells lines, with some heterogeneity in the growth characteristics and histologies. [79] The U87 cell line was originally obtained from a patient with glioblastoma multiforme. [79] Histopathologically, it resembles naturally occurring glioblastoma multiforme, with an infiltrative growth pattern into the brain parenchyma along white matter tracts. [79]

The histology of U87 shows similar characteristics to those glioblastoma multiforme with "nuclear and cytoplasmic atypia and plemorphism". [80] The cell line also shows increased cellularity, mitoses, necrosis, and endothelial proliferation.

SF188 Glioma

The SF188 cell line originally from an 8 year old boy was taken from a biopsy sample directly from the right frontal brain tumor. [81] In 1985, studies described the presence of amplification and expression of the cellular oncogene c-myc in double-minute-containing cells and showed that the amplification is associated with the rearrangement of the c-myc gene in these cells. This finding further contributed to the characterization of this line, but did not provide evidence on c-myc gene amplification in human brain cancer. [80]

060919-Neurosphere Glioma Line

The 060919 glioblastoma neurosphere line was established from resected tissue derived from an adult patient with a temporal lobe glioblastoma, and grown in complete neurosphere media. [82] This line has been shown to be capable of forming invasive, vascular tumors with histological and genetic features consistent with glioblastoma in athymic rodents. It has also exhibited distinctive features of stemlike neurospheres, such as the ability to form self-renewing neurospheres, the potential to differentiate into glial and neuronal lineages, and the expression of high levels of CD133. When intracranially implanted in the brainstem of athymic rats, the 060919 neurosphere line was 100% lethal with histological characteristics of human glioblastoma. These characteristics make this a unique and reproducible brainstem animal model based on a human glioblastoma stem-like neurosphere line. [82]

In Vivo Generation of Intracranial Human Tumor Models

The nude mouse, a non-immunocompetent genetically thymus deficient mouse strain, was introduced by Pantelouris in 1968, allowing heterologous transplantation of human brain tumors to become possible. [3] Since its introduction, human glioma xenografts have allowed for a multitude of investigational studies to be performed. Human tumors, grown in athymic mice, impart certain advantages over murine and rodent models because they maintain many of the biological characteristics of the original biopsy specimens. [116]

Improved transplantation techniques have led to successful, reliable, and reproducible human glioma models in nude mice and athymic rats. Cells kept in culture, or grown as spheroids in vitro, can be implanted either subcutaneously or intracranially. [74, 79] Orthotopic models are attained by stereotactically injecting human glioma cell suspensions into predetermined and localized cortical regions of athymic mouse brains. [3, 78]

Comparison of the Various Human Xenograft Modes

Human glioma xenograft models in nude mice are readily becoming the mainstay of in vivo research because they more closely emulate human gliomas than rodent or murine syngeneic models. [3] Among the cell lines available, both the U87 and the U373 models are popularly used. However, all these models pose several limitations. Each clonal cell line that is maintained in culture does not individually represent the heterogeneous nature of naturally occurring gliomas. [3] In addition, serial subcutaneous passages lead to subsequent alterations in tumor morphologies and frequent sarcomatous degenerations. [3]

Also, xenograft transplantation into nude mice leads to specific interactions between host and graft tissue that alter tumor behavior, but that are not relevant to human tumors. For example, the lack of vascular proliferation seen in these xenograft models is postulated to be due to the murine origin of vascular tissue. [3, 74] These characteristics continue to necessitate the development of improved modeling techniques.

Relevant Studies

Since human glioma cell lines more closely resemble the actual disease entity in humans, xenograft models in nude mice are widely used in experimental neuro-oncology for a variety of studies. Current studies use these models to better understand glioma biology and molecular progression, [83] angiogenesis, [84, 85] mechanisms of drug resistance, [86] and the mode of action and response to various therapeutic agents and treatment modalities. [87]

The use of nude mouse human glioma xenografts has significantly increased our understanding of high grade gliomas.

The use of primary human lines from primary human gliomas, both in monolayer and neurospheres, has paved the way for a deeper understanding of the genetics of malignant glioma. In particular, these models have made it possible to investigate the epigenetic mechanisms of gene expression to define its biological signature. The primary human cell lines maintain the intrinsic heterogeneity of glioblastoma, and are cultured in vitro for an equal length of time. Being, as a result, genetically similar to human glioblastoma, primary human cell lines have been successfully employed in the pursuit of new pathways involved in the differentiation and self-renewal of glioma cells, such as the DNER and EphA2 pathways. Moreover, in consideration of the increasing consensus over the existence of, not one, but many different subtypes of glioblastoma, characterized by different patterns of mutation, it follows that the use of human primary cell lines can be of considerable help to clarify the different subtypes of glioblastoma as well as their varied responsiveness to standard chemotherapeutic agents. [88-90]

RODENT - BRAIN STEM TUMOR MODELS

Brainstem tumors comprise 10-20% of all primary central nervous system tumors in the pediatric population. [91] In the past, tumors of the brainstem were considered to be un-resectable due to the high morbidity and mortality associated with manipulation of the brainstem.

Currently, management decisions are based on the location of the brainstem lesion. Focal lesions, especially tectal gliomas, typically are indolent and can result in obstructive hydrocephalus. Treatment is often limited to symptomatic relief of hydrocephalus by shunting or endoscopic third ventriculostomy. Diffuse lesions are malignant and surgical intervention has not been demonstrated to improve outcome. The development of animal models that closely resemble diffuse brainstem gliomas in the pediatric population has been necessary to advance their clinical treatment and therapeutic options.

There are four main types of brainstem tumors: diffuse brainstem gliomas, focal tumors, exophytic tumors, and cervicomedullary tumors. The most common type of brainstem neoplasms is diffuse brainstem gliomas, which comprises 58-75% of all tumors.

Typically, they are greater than 2 cm at presentation and have a poorly demarcated hyperintense appearance on T2-weighted signal with variable Gadolinium signal enhancement. [92, 93] Most are grade III or IV, malignant fibrillary astrocytomas, have a prognosis similar to that of glioblastoma. [11] Focal tumors have well demarcated borders and may be solid or cystic in nature.

There is no infiltration or edema seen on T1 or T2 sequences and enhancement with gadolinium is variable. Histologically, most focal lesions are low grade (I or II) and considered benign although higher grade lesions have been reported. [94] Exophytic tumors arise from the subependymal glia on the dorsal side of the brainstem. They arise from the 4th ventricle and are hypointense on T1 and hyperintense on T2. The majority includes low grade lesions which have a favorable prognosis.

Cervicomedullary brainstem gliomas resemble intramedullary gliomas of the spinal cord. The solid parts of the tumor demonstrate low to intermediate signal intensity.

Treatment of tumors in the brainstem has been limited by the risk of damaging the structures of the brainstem and by the blood brain barrier. Given that surgical resection in many cases is not a viable option, the need for experimental models to test alternative treatment modalities has arisen. Development of animal brainstem tumor models has provided an important method for testing novel therapies and treatment delivery strategies. The design of an “ideal model” for brainstem tumor research takes many factors into consideration in order to closely replicate tumor behavior in humans. Infiltration of the tumor must occur in a predictable timeframe and be highly reproducible.

The tumor location must also be accessible in order to test novel interventions during the therapeutic window. Functional monitoring and testing are also required to correlate clinical changes with tumor infiltration. Finally, the model should be cost-effective so that chemotherapeutic interventions can be tested in a large enough number of animals to adequately power the study. [3]

Rodents, including mice and rats, have been the primary experimental animals used in brainstem tumor models. Stereotactic implantation of 9L gliosarcoma and F98 glioma tumor lines into the brainstem of rats have overcome previous problems seen with injection of tumor cells. [95] Hemiparesis developed in a predictable manner and correlated to histopathological progression. [95]

Nude mice have also been used to study the metastatic potential of human tumors and have a potential use in brainstem tumor models. [40] A brainstem animal model was recently established using a human glioblastoma stem-like neurosphere line derived from resected tissue. [82] This tumor was shown to derive from brainstem invasive and vascular tumors in athymic rats with histological features consistent with glioblastoma and of stem-like characteristics. Prolonged convection-enhanced delivery for local delivery of infusates into rat brainstems has been shown to be safe, feasible, and beneficial to survival when used for the delivery of continuous infusion of chemotherapeutic agents such as BCNU, IL-13PE, Carboplatin. [97]

These models have been used to improve the technical aspect of local delivery and to optimize brainstem diffusion. Comparative studies on the use of flexible, as opposed to rigid, infusion cannulas can overcome the backflow leakage, and the use of multiple, as opposed to single, cannulas can obtain a homogenous distribution of drugs in brainstem tumors. The rodent brainstem tumor model has also been employed to investigate the safety of continuous drug infusions to the brainstem. [97, 98]

RODENT BRAIN METASTATIC MODEL

Patients with brain metastases have a dismal prognosis, with median survival times of 7.1, 4.2 and 2.3, respectively, for prognostic classes I, II and III. [100] The incidence of brain metastases is 10 times higher than that of primary malignant brain tumors. In most cases, brain metastases derive from primary neoplasms localized in the lung, breast, skin, kidney or colon. [99] The lesions associated with these metastases tend to be highly circumscribed and involve little single cell invasion although, as in the case of anaplastic small cell carcinomas, cases of undefined borders and diffuse invasion patterns also exist. [100]

The actual process of metastasis formation, however, is still unclear. Small metastases, especially if localized in the central nervous system, escape the most advanced diagnostic tools, including computerized tomography (CT), magnetic resonance imaging (MRI) and positron emission tomography (PET). [101]

Due to the difficulties associated with detecting metastases in humans, the metastatic process in the brain has been extensively modeled in animals. In these animal models, which usually involve rodents, tumor cells are injected either orthotopically (into the same organ of origin) or ectopically (into the blood stream) and reach the brain to form solid tumors.

Similar to the glioma models, there are two broad categories of intracranial metastatic models: rodent syngeneic models and human-rodent xenotransplantation models. The former involves murine derived cell lines and can be further divided into ectopic or orthotopic groups.

A subgroup of the orthotopic rodent syngeneic model is that of the genetically engineered mouse models (GEMMs), where genetic manipulations cause primary malignancies that later metastasize to the brain, among other organs.

In these syngeneic models, tumors develop in inbred animals with the same genetic background as the tumor cells. The human-rodent xenotransplantation models denote the ectopic or orthotopic transplantation of human cancer tissues or cell lines into immuno-compromised rodents. [102]

Syngeneic Models

B16-F1 Melanoma Cell Line

The B16-F1 cell line was derived in 1954 from the B16 spontaneous ear melanoma cell line of a C57BL/6 mouse. [103] (Injected intravenously in syngeneic mice, the B16-F1 cells exhibited strong tropism to the lungs. [103] Injected intracardially (ICD), they produced rare brain tumors, which were harvested into the B16-B1 cell line.

Repeated intracardiac injections and selection for brain colonization led to the establishment of two neurotropic cell lines: the B16-B7b cells (metastasizing to the meninges) and the B16-B7n cells (metastasizing to the forebrain). Lines B16-B10b and B16-B10-n followed, with nearly every animal achieving brain metastases, exclusively. Of particular note among the other derivatives of B16 is the highly invasive B16-BL6 melanoma cell line that, after intracarotid artery (ICA) injections in syngeneic mice, was shown to home both to the meninges and the ventricles of the brain. [104]

KHT Sarcoma Cell Line

The KHT mouse sarcoma was derived from a C3H mouse. [105] After ICD injections in female C3H/Bi mice, 60–70 % of the animals developed metastatic brain tumors, localized to the cerebrum, brainstem, cerebellum, and the meninges. McCutcheon and colleagues injected KHT sarcoma cells ICD in C3H mice, followed by adoptive immunotherapy treatment with IL-2 and lymphokine activated killer (LAK) cells. The absence, after treatment, of any reduction in intracerebral metastases and of any evidence of lymphocytic infiltration or cytolytic activity in the brain indicated that brain metastases in patients with systemic malignancies may not respond to intravenous (IV) treatment with LAK cells and IL-2. [106]

4T1 Mammary Carcinoma Cell Line

The 4T1 mammary carcinoma cell line is a thioguanine-resistant variant of the 410.4 cell line derived from a spontaneous mammary tumor in a BALB/cfC3H breeding female. [107] The 4T1 cell line was shown to be highly tumorigenic and capable of spontaneously metastasizing to distant organs, such as lymph nodes, blood, liver, lung, bone and brain. [108]

K-1735 Melanoma Cell Line

The K-1735 murine cells were derived from a cutaneous melanoma in a C3H/HeN mouse and have long been employed in the study of intracerebral metastasis. [109] The cells were shown in vitro to be phenotypically heterogeneous, with the different clones varying in growth, melanin production, and chromosome number. [110] The fact that subcutaneous (SC) implantation of K-1375 cells produced amelanotic tumors, while IV injected cells formed melanotic metastasis, with expression of tumor melanocyte-stimulating hormone receptor, proved the relevance of the metastatic environment. [111] This model has inspired the investigation of brain-specific K-1735 metastasis and blood tumor barrier permeability.

EMT-6 Breast Carcinoma Cell Line

EMT6 was derived from a transplantable murine mammary carcinoma of a BALB/cCRGL mouse after implantation of a hyperplastic mammary alveolar nodule. [112] KHJJ, as the tumor line was named, was propagated in BALB/cKa mice 25 times before being adapted to culture and the cell line was named EMT. In 1971, Stanford University derived the EMT6 clonal isolate, which was used to model hormonal, radiation, and chemotherapeutic effects on primary and metastatic breast cancer growth. [112]

Human Xeno-Transplantation Models

MDA-MB-231 Breast Cell Line

The MDA-MB-231 was derived in 1973 from a human breast tumor pleural effusion at the MD Anderson Cancer Center, where several other cancer cell lines such as MDAMB- 157 and MDA-MB-435 were also developed. [112] The process of MDA-MB-231 brain metastasis was successfully detected by Lorger and Felding-Habermann, who showed that Firefly-luciferase-tagged breast carcinoma cell lines (MDA-MB-231 and MDA-MB-231BR, MDA-MB-435, 4T1, MCF-7) could in fact extravasate through the blood brain barrier between Day 3 and Day 5 after ICA injections in mice, and that astrocytic and microglial reactions were also visible near tumor cells localized in the brain microvasculature. [113] Heyn and colleagues visualized single tumor cell spread to the brain through MR imaging performed 5 h after ICA injections of MDA-MB-231BR breast cancer cells pre-labeled with iron oxide particles. [114] Finally, ICD injection of MDA-MB-231 cells in nude mice revealed the potential role that the matrix metalloprotease 2 may play in breast cancer progression to the brain. [115]

MCF-7 Breast Carcinoma Cell Line

The MCF-7 breast carcinoma cell line was established from a pleural effusion in a patient with metastatic breast carcinoma. [116] MCF-7 cells transfected to overexpress Her-2 were transplanted into the cerebrum of athymic rats. Treatment by intracerebral microinfusion (ICM) with trastuzumab increased median survival by 96 %. In this model, ICM proved superior to the systemic delivery of chemotherapy. The list of brain metastasis models has been extended to include other breast carcinoma cell lines (such as ZR75-1, MDA-MB-361, MDA-MB-468 and MA11) as well as lung carcinoma cell lines (such as A549) and melanoma. [117, 118]

Comparison of the Various Human Xenograft Models

The choice of animal brain metastasis model is dependent on the specific research question under investigation. While syngeneic brain metastasis models facilitate the study of the interaction between tumor cells and the host microenvironment, they confine the researcher to the study of mouse metastatic tumor cells. One advantage of these models is that the K-1735 and the B16 cell lines were established from spontaneous tumors in mice, which can make it easier to explore the interaction between tumor cells and the immuno-competent host in detail. [119] Rapid metastatic spread is also advantageous for the study of molecular mechanisms and for assessment of therapeutic strategies. Human xenograft models are comprised of stromal elements of the tumor from the patient, thereby making them particularly amenable to testing new treatment modalities.

GEMM models are used to study the biological function of genes during neoplastic transformation and tumorigenesis, while specific strains are now being used in selective chemo-prevention and chemotherapy trials. [120] While the genetic alterations that make transformation possible, in terms of induction of oncogenes, suppressor genes, and cell signaling, are still unknown, it is uncertain whether these models accurately reflect human disease.

Conclusion

Despite the considerable effort expended in developing representative animal models of brain metastases, we have not established the process of formation of brain metastasis in humans. Current knowledge indicates that the mechanisms responsible for the metastatic process in the brain involve important interactions within the brain vasculature; the actual contribution of normal brain tissue, however, remains uncertain.

A direction for future research should focus on the challenging task of determining the correct animal models that reflect human disease to accurately identify all the mechanisms behind the metastatic process in the brain.

RODENT - INTRAMEDULLARY SPINAL CORD TUMOR MODELS

Intramedullary spinal cord tumors (IMSCTs) are rare lesions. [121] However, given their infiltrative nature, high recurrence rate, and limited treatment options, their prognosis remains poor. [122] Surgical resection remains the standard of care, despite the inability to often achieve gross total resection given the infiltrative nature of IMSCTs and absence of clear surgical planes between cancerous and normal spinal cord tissue. [122] The use of adjunctive radiotherapy and/or chemotherapy also remains controversial. The efficacy of radiotherapy varies significantly between centers and tumor grade. [123] As in intracranial neoplasms, the efficacy of systemically delivered chemotherapy is largely limited by physiological barriers such as the blood brain barrier, blood-cerebrospinal fluid barrier, and the blood tumor barrier. [124]

In an effort to develop and adequately evaluate novel therapeutics for the treatment of IMSCTs, some animal models have been developed. In contrast to the abundant number of intracranial tumor models, a search of the literature reveals three models of IMSCTs. [125]

Each model utilizes a direct intramedullary injection of a tumor cell suspension and characterizes histopathological, functional, and radiographic progression of disease.

Relevant Studies

Investigators developed two IMSCT rodent models including the F98 glioblastoma cell line and the 9L gliosarcoma cell line, typically used as orthotopic intracranial tumor models. [126] Hindlimb function was evaluated using the Basso, Bresnahan and Beattie scale. [126] This model has been used to examine the efficacy of novel local delivery of the chemotherapeutic agent, taxol, via the OncoGel and Paclimer drug delivery systems. [127, 128] Furthermore, the model has been validated through tumor response to external beam radiation and microsurgical resection. [125] The use of two rodent gliomas with distinct histopathology features, disease progression, and paresis-free survival enables the investigator to choose the onset of paresis which will best suit the needs of the experimental design. The small size of the rodent, however, makes it less ideal for tumor radiographic imaging.

Conclusion

There are relatively few intramedullary spinal cord tumor models compared to the abundant number of intracranial models. The recent development of the rabbit and rodent models represents an important step in helping to adequately develop and evaluate novel therapeutics for the treatment of IMSCTs.

The rodent modes should prove ideal in gaining an understanding of the radiographic, histopathological, and functional progression of the disease.

Rodent - Intraosseous Spinal Cord Tumor Models

Approximately 70% of patients with advanced breast or prostate cancer, and 15% of patients with lung cancer, develop bone metastasis. [129] Of these, 15 to 20% will develop metastasis to the vertebra, with a significant proportion presenting with symptoms including pain, compression fractures, anemia or neurologic sequalae due to spinal cord or nerve root compression. Once tumors have metastasized they are usually incurable and less than 20% of patients survive past 5 years. [129]

Although metastatic spinal cord compression is a major cause of morbidity and mortality, a key obstacle in the study of this disease is the lack of reliable, practical, and reproducible animal models to allow for the exploration of new methodologies for treatment. Despite the many attempts to replicate the effects of the disease in spinal metastases, these models have several limitations, including high mortality rates, the use of rare metastatic routes for cell implantation, and wide range of symptom onset. [130] Better models are needed to provide more reproducible assessments of novel therapeutic agents in animals and of their potential efficacy in humans.

Metastatic Cell Lines

The cell lines used to develop rodent models of intraosseous tumors are human breast (MDA-MB-231), human PC-3 prostate cancer cells, CRL-1666 breast adenocarcinoma, renal carcinoma cells, B16-G3.26 murine melanoma, and Walker 256 rat mammary carcinoma cells. [131, 132] The lines that are used more frequently are human breast (MDA-MB-231), derived from a malignant pleural effusion in a patient with breast adenocarcinoma, the human PC-3 prostate cancer cells and the PC-14 lung cancer cells, derived from a patient with pulmonary adenocarcinoma. Histological analysis has demonstrated the propensity of each of these cancer lines to establish themselves within the bone, destroy it, and then infiltrate the vertebral body and extend out into the epidural space to cause spinal cord compression. [131, 132]

Rodent Relevant Studies

One of the first models was the one set up by Ushio et al. They employed a paraspinal injection of tumor cells, characterized by an infrequent mode of spinal metastasis whereby tumor cells invade the spine rather than growing from within the bone of the vertebra. [130, 133] The pathophysiology of this paraspinal injection is completely different from the typical intraosseous metastasis, as the tumor is believed to invade the vertebra from the paraspinal tissue rather than from the vasculature. In addition, the slow onset of the symptoms of paraplegia – from 3 to 4 weeks – limited the efficiency of the model and its subsequent research potential.

In an attempt to address these limitations, Mantha et al. demonstrated a model of spinal metastasis using a trans-abdominal vertebral implant of a breast cancer cell line. [131] Despite the early onset of paresis that occurred within 14 to 16 days post-surgery, the model required a highly trained surgeon to perform the delicate and lengthy procedure of separating the great vessels from the vertebral column. Moreover, their model implied a high rate of complication (11%) and mortality. This was likely due to the complex dissection of the retro-peritoneal space required to obtain adequate exposure to the vertebral column. Strub et al. demonstrated a metastatic renal cell carcinoma model via intracardiac injection of tumor cells using human 786-O/luciferase cells into the left cardiac ventricle of athymic nude mice. [132] Recently, a line has been described as generating spinal metastases after intracardiac injection. This line, called RBC3, was derived from a metastatic breast adenocarcinoma lesion arising in the osseous spine of a rat following intracardiac injection of MDA-231 human breast cancer cells. [134] Zibly et al. have developed another model characterized by intraosseous implantation of cancer cells in the thoraco-lumbar vertebra through a dorsal approach. [135]

Conclusion

The current model for spinal metastatic cancer involves various tumor cell lines including breast adenocarcinoma, prostate cancer, and lung cancer. Recent advances have not yet proved sufficient to develop reproducible and clinically reliable animal models that can substantially increase our understanding of the temporal and spatial patterns and mechanisms of cancer growth in the spine and the mechanism by which the local host environment influences the spread of cancer. In vivo models are needed to assess novel therapies prior to any translation into human clinical trials.

CONCLUSION

Rodent brain tumor models have made a significant contribution to current knowledge on the biological and genetic features of brain tumors as well as on the correlation between these features and patient clinical and prognostic scenarios. Ideal brain tumor models exhibit predictable and reproducible intracranial growth patterns; have histopathological and biochemical resemblance to human malignant gliomas; and are nonimmunogenic. [136] As described above, a number of models are available for the study of the efficacy and toxicity of different therapeutic approaches, ranging from immunotherapy to therapies based on the use of antiangiogenic agents, alkylating agents or proapoptotic molecules. [136] These glioma models are helping to unravel the biology of tumorigenesis and the etiology of human central nervous system tumors.

Notwithstanding their ultimate contribution, the debate on their clinical merits and translational limits is still open. Meta-analysis studies, for example, have shown that there is substantial concordance between experimental and clinical findings on the therapeutic efficacy of Temozolomide, [137] the most clinically effective chemotherapeutic agent in malignant gliomas [138, 139]. Other agents have a mixed record. In experimental gliomas, nitrosoureas has shown a greater degree of variability in efficacy and little benefit to survival.

Similarly, gene therapy has not yet translated successfully into clinical use. However, it is important to note that, while the feasibility of rodent models needs further elucidation, some chemotherapeutic agents, such as Temozolomide, appeared to be less effective at reducing tumor volume in syngeneic models than they were in actual human-derived tumors. [137] Also, traditional murine tumor models result in a high frequency of false positives: rapidly growing murine tumors are cured as slower growing human tumors progress. [140]

Because of the differences between syngeneic and human models as well as among species of rodents, the translational limits of these glioma models simply highlight the importance of careful model selection at the experimental design stage.

The adoption of a hierarchal strategy, based on rigorous outcome measures and continuous improvements in stringency and consistency, is essential to developing an approach that can facilitate the identification and development of anti-cancer agents and, simultaneously, ensure higher concordance between experimental models and successive findings in clinical trials. [137, 140]

REFERENCES

[1] Barth, R. F. A critical assessment of boron neutron capture therapy: an overview. *J. Neurooncol.* 62:1-5, 2003.

[2] Barth, R. F. Rat brain tumor models in experimental neuro-oncology: the 9L, C6, T9, F98, RG2 (D74), RT-2 and CNS-1 gliomas. *J. Neurooncol.* 36:91-102, 1998.

[3] Peterson, D. L., Sheridan, P. J., Brown, W. E., Jr. Animal models for brain tumors: historical perspectives and future directions. *J. Neurosurg.* 80:865-876, 1994.

[4] Grobben, B., De Deyn, P. P., Slegers, H. Rat C6 glioma as experimental model system for the study of glioblastoma growth and invasion. *Cell Tissue Res.* 310:257-270, 2002.

[5] Benda, P., Someda, K., Messer, J., Sweet, W. H. Morphological and immunochemical studies of rat glial tumors and clonal strains propagated in culture. *J. Neurosurg.* 34: 310-323, 1971.

[6] Schmidek, H. H., Nielsen, S. L., Schiller, A. L., Messer, J. Morphological studies of rat brain tumors induced by N-nitrosomethylurea. *J. Neurosurg.* 34:335-340, 1971.

[7] Barker, M., Hoshino, T., Gurcay, O., Wilson, C. B., Nielsen, S. L., Downie, R., et al. Development of an animal brain tumor model and its response to therapy with 1,3-bis (2-chloroethyl)-1-nitrosourea. *Cancer Res.* 33:976-986, 1973.

[8] Ko, L., Koestner, A., Wechsler, W. Morphological characterization of nitrosourea-induced glioma cell lines and clones. *Acta Neuropathol.* 51:23-31, 1980.

[9] Kobayashi, N., Allen, N., Clendenon, N. R., Ko, L. W. An improved rat brain-tumor model. *J. Neurosurg.* 53:808-815, 1980.

[10] Tzeng, J. J., Barth, R. F., Orosz, C. G., James, S. M. Phenotype and functional activity of tumor-infiltrating lymphocytes isolated from immunogenic and nonimmunogenic rat brain tumors. *Cancer Res.* 51:2373-2378, 1991.

[11] Albright, A. L., Guthkelch, A. N., Packer, R. J., Price, R. A., Rourke, L. B. Prognostic factors in pediatric brain-stem gliomas. *J. Neurosurg.* 65:751-755, 1986.

[12] Rhines, L. D., Sampath, P., Dolan, M. E., Tyler, B. M., Brem, H., Weingart, J. O6-benzylguanine potentiates the antitumor effect of locally delivered carmustine against an intracranial rat glioma. *Cancer Res*. 60(22):6307-10, 2000.
[13] Kishida, Y., Natsume, A., Toda, H., Toi, Y., Motomura, K., Koyama, H., Matsuda, K., Nakayama, O., Sato, M., Suzuki, M., Kondo, Y., Wakabayashi, T. Correlation between Quantified Promoter Methylation and Enzymatic Activity of O6-Methylguanine-DNA Methyltransferase in Glioblastomas. *Tumour Biology*. Vol. 33(No. 2):373–381, 2012.
[14] Gerson, S. L. Clinical Relevance of MGMT in the Treatment of Cancer. *J. Clin. Oncol.* 20: 2388-2399, 2002.
[15] Aas, A. T., Brun, A., Blennow, C., Stromblad, S., Salford, L. G. The RG2 rat glioma model. *J. Neurooncol*. 23:175-183, 1995.
[16] Kruse, C. A., Molleston, M. C., Parks, E. P., Schiltz, P. M., Kleinschmidt-DeMasters, B. K., Hickey, W. F. A rat glioma model, CNS-1, with invasive characteristics similar to those of human gliomas: A comparison to 9L gliosarcoma. *J. Neurooncol*. 22:191-200, 1994.
[17] Doblas, S., He, T., Saunders, D., Pearson, J., Hoyle, J., Smith, N., et al. Glioma morphology and tumor-induced vascular alterations revealed in seven rodent glioma models by in vivo magnetic resonance imaging and angiography. *J. Magn. Reson. Imaging*. Aug.;32(2):267-75, 2010.
[18] Pacholska, A., Wirth, T., Samaranayake, H., Pikkarainen, J., Ahmad, F., and Yla-Herttuala, S. Increased invasion of malignant gliomas after 15-LO-1 and HSV-tk/ ganciclovir combination gene therapy. *Cancer Gene Ther.* Dec.;19(12):870-4, 2012.
[19] Siesjo, P., Visse, E., Lindvall, M., Salford, L., Sjogren, H. O. Immunization with mutagen-treated (tum-) cells causes rejection of nonimmunogenic rat glioma isografts. *Cancer Immunol. Immunother*. 37: 67–74,1993.
[20] Bexell, D., Gunnarsson, S., Siesj, P., Bengzon, J., Darabi, A. CD133+ and nestin+ tumor-initiating cells dominate in N29 and N32 experimental gliomas. *Int. J. Cancer* 125:15-22, 2009.
[21] Kida, Y., Cravioto, H., Hochwald, G. M., et al. Immunity to transplantable nitrosourea induced neurogenic tumors. II. Immunoprophylaxis of tumors of the brain. *Journal of neuropathology and experimental neurology,* Vol. 42, pp. 122– 135, 1983.
[22] Jeffes, E. W., Zhang, J. G., Hoa, N., et al. Antiangiogenic drugs synergize with a membrane macrophage colony-stimulating factor-based tumor vaccine to therapeutically treat rats with an established malignant glioma. *Immunol*. Mar. 1;174(5): 2533-43, (2005).
[23] Pietronigro, D., Drnovsky, F., Cravioto, H., and Ransohoff, J. DTI-015 produces cures in T9 gliosarcoma. *Neoplasia* Vol. 5, pp. 17–22, 2003.
[24] Copeland, D. D., Talley, F. A. and Bigner, D. D. The fine structure of intracranial neoplasms induced by the inoculation of avian sarcoma virus in neonatal and adult rats. *American Journal Pathology*, Vol. 83, pp. 149–176, 1976.
[25] Valerie, K., Hawkins, W., Farnsworth, J., et al. Substantially improved in vivo radiosensitization of rat glioma with mutant HSV-TK and acyclovir. *Cancer Gene Ther.* Jan.;8(1):3-8, 2001.
[26] Mourad, P. D., Farrell, L., Stamps, L. D., et al. Quantitative assessment of glioblastoma invasion in vivo. *Cancer Letters*, Vol. 192, pp. 97–107, 2003.

[27] Fross, R. D., Warnke, P. C., Groothuis, D. R. Blood flow and blood-to-tissue transport in 9L gliosarcomas: the role of the brain tumor model in drug delivery research. *J. Neurooncol.* 11:185-197, 1991.

[28] Guerin, C., Laterra, J., Drewes, L. R., Brem, H., Goldstein, G. W. Vascular expression of glucose transporter in experimental brain neoplasms. *Am. J. Pathol.* 140:417-425, 1992.

[29] Rajan, S. S., Rosa, L., Francisco, J., Muraki, A., Carvlin, M., Tuturea, E. MRI characterization of 9L-glioma in rat brain at 4.7 Tesla. *Magn. Reson. Imaging* 8:185-190, 1990.

[30] Yoshida, T., Shimizu, K., Ushio, Y., Hayakawa, T., Mogami, H., Sakamoto, Y. The mechanism and overcoming of resistance in ACNU-resistant sublines of C6 and 9L rat glioma. *J. Neurooncol.* 5:195-203, 1987.

[31] Sills, A. K., Williams, J. I., Tyler, B. M., Epstein, D. S., Sipos, E. P., Davis, J. D., McLane, M. P., Pitchford, S., Chesire, K., Gannon, F. H., Kinney, W. A., Chao, T. L, Donowitz, M., Laterra, J., Zasloff, M., Brem, H. Squalamine Inhibits Angiogenesis And Tumor Growth In Vivo And Perturbs Embryonic Vasculature. *Cancer Research*, 58: 2784-2792, 1998.

[32] Tyler, B., Fowers, K. D., Li, K. W., Recinos, V. R., Caplan, J. M., Hdeib, A., et al. A thermal gel depot for local delivery of paclitaxel to treat experimental brain tumors in rats. *J. Neurosurg.* Aug.;113(2):210-7, 2010.

[33] Li, K. W., Dang, W., Tyler, B. M., Troiano, G., Tihan, T., Brem, H., et al. Polilactofate microspheres for Paclitaxel delivery to central nervous system malignancies. *Clin. Cancer Res.* 9:3441-3447, 2003.

[34] Henderson, S. D., Kimler, B. F., Morantz, R. A. Radiation therapy of 9L rat brain tumors. *Int. J. Radiat. Oncol. Biol. Phys.* 7:497-502, 1981.

[35] DiMeco, F., Rhines, L. D., Hanes, J., Tyler, B. M., Brat, D., Torchiana, E., et al. Paracrine delivery of IL-12 against intracranial 9L gliosarcoma in rats. *J. Neurosurg.* 92: 419-427, 2000.

[36] Hsu, W., Lesniak, M. S., Tyler, B., Brem, H. Local delivery of interleukin-2 and adriamycin is synergistic in the treatment of experimental malignant glioma. *J. Neurooncol.* 74:135-140, 2005.

[37] Eis, M., Els, T., Hoehn-Berlage, M. High resolution quantitative relaxation and diffusion MRI of three different experimental brain tumors in rat. *Magn. Reson. Med.* 34:835-844, 1995.

[38] Olivi, A., Ewend, M. G., Utsuki, T., Tyler, B., Domb, A. J., Brat, D. J., et al. Interstitial delivery of carboplatin via biodegradable polymers is effective against experimental glioma in the rat. *Cancer Chemother. Pharmacol.* 39:90-96, 1996.

[39] Pradilla, G., Thai, Q. A., Legnani, F. G., Clatterbuck, R. E., Gailloud, P., Murphy, K. J., et al. Local delivery of ibuprofen via controlled-release polymers prevents angiographic vasospasm in a monkey model of subarachnoid hemorrhage. *Neurosurgery.* 2005 Jul.;57(1 Suppl.):184-90; discussion 184-90.

[40] Chicoine, M. R., Silbergeld, D. L. Invading C6 glioma cells maintaining tumorigenicity. *J. Neurosurg.* 83:665-671, 1995.

[41] Resnicoff, M., Li, W., Basak, S., Herlyn, D., Baserga, R., Rubin, R. Inhibition of rat C6 glioblastoma tumor growth by expression of insulin-like growth factor I receptor antisense mRNA. *Cancer Immunol. Immunother.* 42:64-68, 1996.

[42] Farrell, C. L., Stewart, P. A., Del Maestro, R. F. A new glioma model in rat: the C6 spheroid implantation technique permeability and vascular characterization. *J. Neurooncol*. 4:403-415, 1987.

[43] Groothuis, D. R., Pasternak, J. F., Fischer, J. M., Blasberg, R. G., Bigner, D. D., Vick, N. A. Regional measurements of blood flow in experimental RG-2 rat gliomas. *Cancer Res*. 43:3362-3367, 1983.

[44] Molnar, P., Lapin, G. D., Groothuis, D. R. The effects of dexamethasone on experimental brain tumors: I. Transcapillary transport and blood flow in RG-2 rat gliomas. *J. Neurooncol*. 25:19-28, 1995.

[45] Vallbo, C., Bergenheim, T., Hedman, H., et al. The antimicrotubule drug estramustine but not irradiation induces apoptosis in malignant glioma involving AKT and caspase pathways. *Journal of neuro-oncology*, Vol. 56, pp. 143–148, 2002.

[46] Andersson, U., Grankvist, K., Bergenheim, A. T., et al. Rapid induction of long-lasting drug efflux activity in brain vascular endothelial cells but not malignant glioma following irradiation. *Medical oncology*, Vol. 19, pp. 1–9, 2002.

[47] Griffin, J. L., Lehtimaki, K. K., Valonen, P. K., et al. Assignment of 1H nuclear magneticresonance visible polyunsaturated fatty acids in BT4C gliomas undergoing ganciclovir-thymidine kinase gene therapy-induced programmed cell death. *Cancer Research*, Vol. 63, pp. 3195–3201, 2003.

[48] Harada, K., Yoshida, J., Mizuno, M., et al. Growth inhibition of intracerebral rat glioma by transfection induced human interferon-beta. *Journal of Surgical Oncology*, Vol. 59, pp. 105–109, 1995.

[49] Dai, C., Holland, E. C. Glioma models. *Biochim. Biophys. Acta* 1551:M19-27, 2001.

[50] Ausman, J. I., Shapiro, W. R., Rall, D. P. Studies on the chemotherapy of experimental brain tumors: development of an experimental model. *Cancer Res*. 30:2394-2400, 1970.

[51] Zagzag, D., Esencay, M., Mendez, O., Yee, H., Smirnova, I., Huang, Y., Chiriboga, L., Lukyanov, E., Liu, M., Newcomb, E. Hypoxia- and Vascular Endothelial Growth Factor-Induced Stromal Cell-Derived Factor-1α/CXCR4 Expression in Glioblastomas. One Plausible Explanation of Scherer's Structures. *Am. J. Pathol*. August; 173(2): 545–560, 2008.

[52] Jacobs, V. L., Valdes, P. A., Hickey, W. F., De Leo, J. A. Current review of in vivo GBM rodent models: Emphasis on the CNS-1 tumour model. *ASN Neuro*. Aug. 3;3(3):e 00063, 2011.

[53] Desaknai, S., Lumniczky, K., Esik, O., Hamada, H., Safrany, G. Local tumour irradiation enhances the anti-tumour effect of a double-suicide gene therapy system in a murine glioma model. *J. Gene Med*. 5:377-385, 2003.

[54] Prins, R. M., Odesa, S. K., Liau, L. M. Immunotherapeutic targeting of shared melanoma-associated antigens in a murine glioma model. *Cancer Res*. 63:8487-8491, 2003.

[55] Maes, W., Van Gool, S. W. Experimental immunotherapy for malignant glioma: lessons from two decades of research in the GL261 model. *Cancer Immunol. Immunother*. 60:153–160, 2011.

[56] Ohgaki, H., Kleihues, P. Genetic pathways to primary and secondary glioblastoma. *Am. J. Pathol.* 170:1445-1453, 2007.

[57] Watanabe, K., Tachibana, O., Sata, K., Yonekawa, Y., Kleihues, P., Ohgaki, H. Overexpression of the EGF receptor and p53 mutations are mutually exclusive in the

evolution of primary and secondary glioblastomas. *Brain Pathol.* 6:217-223; discussion 223-214, 1996.

[58] Cancer Genome Atlas Research Network. Comprehensive genomic characterization defines human glioblastoma genes and core pathways. *Nature* 455:1061-1068, 2008.

[59] Holland, E. C., Hively, W. P., DePinho, R. A., Varmus, H. E. A constitutively active epidermal growth factor receptor cooperates with disruption of G1 cell-cycle arrest pathways to induce glioma-like lesions in mice. *Genes Dev.* 12:3675-3685, 1998.

[60] Tohma, Y., Gratas, C., Biernat, W., Peraud, A., Fukuda, M., Yonekawa, Y., et al. PTEN (MMAC1) mutations are frequent in primary glioblastomas (de novo) but not in secondary glioblastomas. *J. Neuropathol. Exp. Neurol.* 57:684-689, 1998.

[61] Parsons, D. W., Jones, S., Zhang, X., Lin, J. C., Leary, R. J., Angenendt, P., et al. An integrated genomic analysis of human glioblastoma multiforme. *Science* 321:1807-1812, 2008.

[62] Verhaak, R. G., Hoadley, K. A., Purdom, E., Wang, V., Qi, Y., Wilkerson, M. D., et al. Integrated genomic analysis identifies clinically relevant subtypes of glioblastoma characterized by abnormalities in PDGFRA, IDH1, EGFR, and NF1. *Cancer Cell* 17: 98-110, 2010.

[63] Guerrero-Cazares, H., Chaichana, K. L., Quinones-Hinojosa, A. Neurosphere culture and human organotypic model to evaluate brain tumor stem cells. *Methods Mol. Biol.* 568:73-83, 2009.

[64] Moreadith, R. W., Radford, N. B. Gene targeting in embryonic stem cells: the new physiology and metabolism. *J. Mol. Med.* 75:208-216, 1997.

[65] Danks, R. A., Orian, J. M., Gonzales, M. F., Tan, S. S., Alexander, B., Mikoshiba, K., et al. Transformation of astrocytes in transgenic mice expressing SV40 T antigen under the transcriptional control of the glial fibrillary acidic protein promoter. *Cancer Res.* 55: 4302-4310, 1995.

[66] Jacks, T., Remington, L., Williams, B. O., Schmitt, E. M., Halachmi, S., Bronson, R. T., et al. Tumor spectrum analysis in p53-mutant mice. *Curr. Biol.* 4:1-7, 1994.

[67] Kwon, C. H., Zhao, D., Chen, J., Alcantara, S., Li, Y., Burns, D. K., et al. Pten haploinsufficiency accelerates formation of high-grade astrocytomas. *Cancer Res.* 68: 3286-3294, 2008.

[68] Akagi, K., Sandig, V., Vooijs, M., Van der Valk, M., Giovannini, M., Strauss, M., et al. Cre-mediated somatic site-specific recombination in mice. *Nucleic Acids Res.* 25:1766-1773, 1997.

[69] Alcantara Llaguno, S., Chen, J., Kwon, C. H., Jackson, E. L., Li, Y., Burns, D. K., et al. Malignant astrocytomas originate from neural stem/progenitor cells in a somatic tumor suppressor mouse model. *Cancer Cell* 15:45-56, 2009.

[70] Marumoto, T., Tashiro, A., Friedmann-Morvinski, D., Scadeng, M., Soda, Y., Gage, F. H., et al. Development of a novel mouse glioma model using lentiviral vectors. *Nat. Med.* 15:110-116, 2009.

[71] Holland, E. C., Varmus, H. E. Basic fibroblast growth factor induces cell migration and proliferation after glia-specific gene transfer in mice. *Proc. Natl. Acad. Sci. US* 95: 1218-1223, 1998.

[72] De Vries, N. A., Bruggeman, S. W., Hulsman, D., de Vries, H. I., Zevenhoven, J., Buckle, T., et al. Rapid and robust transgenic high-grade glioma mouse models for therapy intervention studies. *Clin. Cancer Res.* 16: 3431-3441, 2010.

[73] Shiras, A., Chettiar, S. T., Shepal, V., Rajendran, G., Prasad, G. R., Shastry, P. Spontaneous transformation of human adult nontumorigenic stem cells to cancer stem cells is driven by genomic instability in a human model of glioblastoma. *Stem Cells* 25: 1478-1489, 2007.

[74] Goldbrunner, R. H., Wagner, S., Roosen, K., Tonn, J. C. Models for assessment of angiogenesis in gliomas. *J. Neurooncol.* 50:53-62, 2000.

[75] Guo, M., Roman, R. J., Falck, J. R., Edwards, P. A., Scicli, A. G. Human U251 glioma cell proliferation is suppressed by HET0016 [N-hydroxy-N′-(4-butyl-2-methylphenyl) formamidine], a selective inhibitor of CYP4A. *J. Pharmacol. Exp. Ther.* 315(2):526–533, 2005.

[76] Qiang, L., Yang, Y., Ma, Y. J., Chen, F. H., Zhang, L. B., Liu, W., Qi, Q., Lu, N., Tao, L., Wang, X. T., You, Q. D., Guo, Q. L. Isolation and characterization of cancer stem like cells in human glioblastoma cell lines. *Cancer Lett.* Jun. 28;279(1):13-21, 2009.

[77] Ponten, J., Macintyre, E. H. Long term culture of normal and neoplastic human glia. *Acta Pathol. Microbiol. Scand.* 74:465-486, 1968.

[78] Branle, F., Lefranc, F., Camby, I., Jeuken, J., Geurts-Moespot, A., Sprenger, S., et al. Evaluation of the efficiency of chemotherapy in in vivo orthotopic models of human glioma cells with and without 1p19q deletions and in C6 rat orthotopic allografts serving for the evaluation of surgery combined with chemotherapy. *Cancer* 95:641-655, 2002.

[79] Ozawa, T., Wang, J., Hu, L. J., Bollen, A. W., Lamborn, K. R., Deen, D. F. Growth of human glioblastomas as xenografts in the brains of athymic rats. *In Vivo* 16:55-60, 2002.

[80] Trent, J., Meltzer, P., Rosenblum, M., Harsh, G., Kinzler, K., et al. Evidence for rearrangement, amplification, and expression of c-myc in a human glioblastoma. *Proc. Natl. Acad. Sci. US* 83:470–473206, 1986.

[81] Vladimirova, V., Mikeska, T., Waha, A., Soerensen, N., Xu, J., Reynolds, P. C., Pietsch, T. Aberrant methylation and reduced expression of LHX9 in malignant gliomas of childhood. *Neoplasia* 11:700–711, 2009.

[82] Siu, I. M., Tyler, B. M., Chen, J. X., Eberhart, C. G., Thomale, U. W., et al. Establishment of a human glioblastoma stemlike brainstem rodent tumor model. *J. Neurosurg. Pediatr.* 6: 92–97, 2010.

[83] Chan, J. A., Krichevsky, A. M., Kosik, K. S. MicroRNA-21 is an antiapoptotic factor in human glioblastoma cells. *Cancer Res.* 65:6029-6033, 2005.

[84] Ma, J., Fei, Z. L., Klein-Szanto, A., Gallo, J. M. Modulation of angiogenesis by human glioma xenograft models that differentially express vascular endothelial growth factor. *Clin. Exp. Metastasis* 16:559-568, 1998.

[85] Stan, A. C., Nemati, M. N., Pietsch, T., Walter, G. F., Dietz, H. In vivo inhibition of angiogenesis and growth of the human U-87 malignant glial tumor by treatment with an antibody against basic fibroblast growth factor. *J. Neurosurg.* 82:1044-1052, 1995.

[86] Lund, E. L., Hansen, L. T., Kristjansen, P. E. Augmenting tumor sensitivity to topotecan by transient hypoxia. *Cancer Chemother. Pharmacol.* 56:473-480, 2005.

[87] Lal, B., Xia, S., Abounader, R., Laterra, J. Targeting the c-Met pathway potentiates glioblastoma responses to gamma-radiation. *Clin. Cancer Res.* 11:4479-4486, 2005.

[88] Sun, P., Xia, S., Lal, B., et al. DNER, an epigenetically modulated gene, regulates glioblastoma-derived neurosphere cell differentiation and tumor propagation. *Stem Cells* 27:1473–1486, 2009.

[89] Binda, E., Visioli, A., Giani, F., Lamorte, G., Copetti, M., et al. The EphA2 receptor drives self-renewal and tumorigenicity in stem-like tumor-propagating cells from human glioblastomas. *Cancer Cell* 22(6): 765-780, 2012.

[90] Beier, D., Hau, P., Proescholdt, M., Lohmeier, A., Wischhusen, J., et al. CD133(+) and CD133(−) glioblastoma-derived cancer stem cells show differential growth characteristics and molecular profiles. *Cancer Res.* 67:4010–4015, 2007.

[91] Jallo, G. I., Freed, D., Roonprapunt, C., Epstein, F. Current management of brainstem gliomas. *Annals of Neurosurgery* 3:1-17, 2003.

[92] Fischbein, N. J., Prados, M. D., Wara, W., Russo, C., Edwards, M. S., Barkovich, A. J. Radiologic classification of brain stem tumors: correlation of magnetic resonance imaging appearance with clinical outcome. *Pediatr. Neurosurg.* 24:9-23, 1996.

[93] Jallo, G. I., Biser-Rohrbaugh, A., Freed, D. Brainstem gliomas. *Childs Nerv. Syst.* 20: 143-153, 2004.

[94] Abbott, R. Tumors of the medulla. *Neurosurg. Clin. N. Am.* 4:519-527, 1993.

[95] Jallo, G. I., Volkov, A., Wong, C., Carson, B. S., Sr., Penno, M. B. A novel brainstem tumor model: functional and histopathological characterization. *Childs Nerv. Syst.* 22: 1519-1525, 2006.

[96] Sandberg, D. I., Edgar, M. A., Souweidane, M. M. Convection-enhanced delivery into the rat brainstem. *J. Neurosurg.* 96:885-891, 2002.

[97] Thomale, U. W., Tyler, B., Renard, V. M., Dorfman, B., Guarnieri, M., Haberl, H. E., Jallo, G. I. Local chemotherapy in the rat brainstem with multiple catheters: a feasibility study. *Childs Nerv. Syst.* Jan.;25(1):21-8, 2008.

[98] Thomale, U. W., Tyler, B., Renard, V., Dorfmann, B., Chacko, V. P., Carson, B. S., Haberl, E. J., Jallo, G. I. Neurological grading, survival, MR imaging and histological evaluation in the rat brainstem glioma model. *Childs Nerv. Syst.* Apr.;25(4):433-41, 2009.

[99] Barnholtz-Sloan, J. S. Incidence proportions of brain metastases in patients diagnosed (1973 to 2001) in the Metropolitan Detroit Cancer Surveillance System. *J. Clin. Oncol.* 22: 2865–2872, 2004.

[100] Tosoni, A., Ermani, M., Brandes, A. A. The pathogenesis and treatment of brain metastases: a comprehensive review. *Crit. Rev. Oncol./Hematol.* 52:199–215, 2004.

[101] Chambers, A. F., Groom, A. C., MacDonald, I. C. Dissemination and growth of cancer cells in metastatic sites. *Nat. Rev. Cancer* 2:563–572, 2002.

[102] Daphu, I., Sundstrøm, T., Horn, S., Huszthy, P. C., Niclou, S. P., Sakariassen, P. Ø., Immervoll, H., Miletic, H., Bjerkvig, R., Thorsen, F. In vivo animal models for studying brain metastasis: value and limitations. *Clin. Exp. Metastasis* 30(5):695-710, 2013.

[103] Fidler, I. J. Selection of successive tumour lines for metastasis. *Nat. New Biol.* 242: 148–149, 1973.

[104] Schackert, G., Fidler, I. J. Site-specific metastasis of mouse melanomas and a fibrosarcoma in the brain or meninges of syngeneic animals. *Cancer Res.* 48:3478–3484, 1988.

[105] Kallman, R. F., Silini, G., Van Putten, L. M. Factors influencing the quantitative estimation of the in vivo survival of cells from solid tumors. *J. Natl. Cancer Inst.* 39: 539–549, 1967.

[106] Conley, F. K. Development of a metastatic brain tumor model in mice. *Cancer Res.* 39: 1001–1007, 1979.

[107] Yamashina, K., Heppner, G. H. Correlation of frequency of inducedmutation andmetastatic potential in tumor cell lines from a single mouse mammary tumor. *Cancer Res.* 45:4015–4019, 2006.

[108] Lelekakis, M., Moseley, J. M., Martin, T. J., Hards, D., Williams, E., Ho, P., Lowen, D., Javni, J., Miller, F. R., Slavin, J., Anderson, R. L. A novel orthotopic model of breast cancer metastasis to bone. *Clin. Exp. Metastasis* 17:163–170, 1999.

[109] Kripke, M. L. Speculations on the role of ultraviolet radiation in the development of malignant melanoma. *J. Natl. Cancer Inst.* 63:541–548, 1979.

[110] Fidler, I. J., Gruys, E., Cifone, M. A., Barnes, Z., Bucana, C. Demonstration of multiple phenotypic diversity in a murine melanoma of recent origin. *J. Natl. Cancer Inst.* 67: 947–956, 1981.

[111] Radinsky, R., Beltran, P. J., Tsan, R., Zhang, R., Cone, R. D., Fidler, I. J. Transcriptional induction of the melanocyte-stimulating hormone receptor in brain metastases of murine K-1735 melanoma. *Cancer Res.* 55:141–148, 1995.

[112] Gorczynski, R. M., Chen, Z., Diao, J., Khatri, I., Wong, K., et al. Breast cancer cell CD200 expression regulates immune response to EMT6 tumor cells in mice. *Breast Cancer Res. Treat.* 123: 405–415, 2010.

[113] Lorger, M., Felding-Habermann, B. Capturing changes in the brain microenvironment during initial steps of breast cancer brain metastasis. *Am. J. Pathol.* 176:2958–2971, 2010.

[114] Heyn, C., Ronald, J. A., Ramadan, S. S., Snir, J. A., Barry, A. M., Mac- Kenzie, L. T., Mikulis, D. J., Palmieri, D., Bronder, J. L., Steeg, P. S., Yoneda, T., MacDonald, I. C., Chambers, A. F., Rutt, B. K., Foster, P. J. In vivo MRI of cancer cell fate at the single-cell level in a mouse model of breast cancer metastasis to the brain. *Magn. Reson. Med.* 56:1001–1010, 2006.

[115] Tester, A. M., Waltham, M., Oh, S.-J., Bae, S. N., Bills, M. M., Walker, E. C., Kern, F. G., Stetler-Stevenson, W. G., Lippman, M. E., Thompson, E. W. Pro-matrix metallo-proteinase-2 transfection increases orthotopic primary growth and experimental metastasis of MDA-MB-231 human breast cancer cells in nude mice. *Cancer Res.* 64: 652–658, 2004.

[116] Soule, H. D., Vazguez, J., Long, A., Albert, S., Brennan, M. A human cell line from a pleural effusion derived from a breast carcinoma. *J. Natl. Cancer Inst.* 51:1409–1416, 1973.

[117] Engel, L. W., Young, N. A., Tralka, T. S., Lippman, M. E., O0Brien, S. J., Joyce, M. J. Establishment and characterization of three new continuous cell lines derived from human breast carcinomas. *Cancer Res.* 38:3352–33, 1978.

[118] Lieber, M., Smith, B., Szakal, A., Nelson-Rees, W., Todaro, G. (1976) A continuous tumor-cell line from a human lung carcinoma with properties of type II alveolar epithelial cells. *Int. J. Cancer* 17:62–70, 1976.

[119] Fidler, I. J. The role of the organ microenvironment in brain metastasis. *Semin. Cancer Biol.* 21:107–112, 2011.

[120] Wagner, K. U. Models of breast cancer: quo vadis, animal modeling? *Breast Cancer Res.* 6:31–38, 2004.

[121] Balmaceda, C. Chemotherapy for intramedullary spinal cord tumors. *J. Neurooncol.* 47: 293-307, 2000.

[122] Bowers, D. C., Weprin, B. E. Intramedullary Spinal Cord Tumors. *Curr. Treat. Options Neurol.* 5:207-212, 2003.

[123] Jallo, G. I., Danish, S., Velasquez, L., Epstein, F. Intramedullary low-grade astrocytomas: long-term outcome following radical surgery. *J. Neurooncol.* 53:61-66, 2001.

[124] Raza, S. M., Pradilla, G., Legnani, F. G., Thai, Q. A., Olivi, A., Weingart, J. D., et al. Local delivery of antineoplastic agents by controlled-release polymers for the treatment of malignant brain tumours. *Expert Opin. Biol. Ther.* 5:477-494, 2005.

[125] Pennant, W. A., Sciubba, D. M., Noggle, J. C., Tyler, B. M., Tamargo, R. J., Jallo, G. I. Microsurgical removal of intramedullary spinal cord gliomas in a rat spinal cord decreases onset to paresis, an animal model for intramedullary tumor treatment. *Childs Nerv. Syst.* 24:901-907, 2008.

[126] Caplan, J., Pradilla, G., Hdeib, A., Tyler, B. M., Legnani, F. G., Bagley, C. A., Brem, H., Jallo, G. A novel model of intramedullary spinal cord tumors in rats: functional progression and histopathological characterization. *Neurosurgery*. 2006;59:193–200. discussion 193–200.

[127] Gok, B., McGirt, M. J., Sciubba, D. M., Garces-Ambrossi, G., Nelson, C., Noggle, J., et al. Adjuvant treatment with locally delivered OncoGel delays the onset of paresis after surgical resection of experimental spinal column metastasis. *Neurosurgery* 65:193-199; discussion 199-200, 2009.

[128] Tyler, B., Hdeib, A., Caplan, J., Legnani, F. G., Fowers, K. D., Brem, H., Jallo, G., Pradilla, G. Delayed onset of paresis in rats with experimental intramedullary spinal cord gliosarcoma following intratumoral administration of the paclitaxel delivery system OncoGel. *J. Neurosurgery: Spine*, Jan.;16(1): 93-101, 2012.

[129] Coleman, R. E. Metastatic bone disease: clinical features, pathophysiology and treatment strategies. *Cancer Treat. Rev.* 27(3):165-176. 130. 2001.

[130] Ushio, Y., Posner, R., Kim, J. H., Shapiro, W. R., Posner, J. B.: Treatment of experimental spinal cord compression caused by extradural neoplasms. *J. Neurosurg.* 1977, 47(3):380-390.

[131] Mantha, A., Legnani, F. G., Bagley, C. A., Gallia, G. L., Garonzik, I., Pradilla, G., Amundson, E., Tyler, B. M., Brem, H., Gokaslan, Z. L. A novel rat model for the study of intraosseous metastatic spine cancer. *J. Neurosurg. Spine* Mar.;2(3):303-7, 2005.

[132] Strube, A., Stepina, E., Mumberg, D., Scholz, A., Hauff, P., Kakonen, S. M. Characterization of a new renal cell carcinoma bone metastasis mouse model. *Clin. Exp. Metastasis* 27(5):319-330, 2010.

[133] Ushio, Y., Posner, R., Posner, J. B., Shapiro, W. R. Experimental spinal cord compression by epidural neoplasm. *Neurology* 27(5):422-429, 1977.

[134] Zadnik, P., Sarabia-Estrada, R., Groves, M., Molina, C., Jackson, C., McCarthy, E., Gokaslan, Z. L., Bydon, A., Wolinsky, J. P., Witham, T. F., Sciubba, D. M. A novel animal model of human breast metastasis tothe spine: A pilot study using intracardiac injection and luciferase expressing cells. *J. Neurosurg. Spine*. 18(3):217-25, 2013.

[135] Zibly, Z., Schlaff, C. D., Gordon, I., Munasinghe, J., Camphausen, K. A. A novel rodent model of spinal metastasis and spinal cord compression *BMC Neuroscience* 13: 137, 2012.

[136] Sun, L. *Animal Models of Glioma- Glioma-Exploring its Biology and Practical Relevance,* book edited by Anirban Ghosh, ISBN 978-953-307-379-8, Published: November 2, 2011.

[137] Hirst, T. C., Vesterinen, H. M., Sena, E. S., Egan, K. J., Macleod, M. R., Whittle, I. R. Systematic review and meta-analysis of temozolomide in animal models of glioma: was clinical efficacy predicted? *Br. J. Cancer* 108(1):64-71, 2013.

[138] Stupp, R., Mason, W. P., van den Bent, M. J., Weller, M., Fisher, B., Taphoorn, M. J., Belanger, K., Brandes, A. A., Marosi, C., Bogdahn, U., Curschmann, J., Janzer, R. C., Ludwin, S. K., Gorlia, T., Allgeier, A., Lacombe, D., Cairncross, J. G., Eisenhauer, E., Mirimanoff, R. O. Radiotherapy plus concomitant and adjuvant temozolomide for glioblastoma. *N. Engl. J. Med*. 352(10): 987–996, 2005.

[139] Van den Bent, M. J., Hegi, M. E., Stupp, R. Recent developments in the use of chemotherapy in brain tumours. *Eur. J. Cancer* 42(5): 582–588, 2006.

[140] Talmadge, J. E., Singh, R. K., Fidler, I. J., Raz, A. Murine models to evaluate novel and conventional therapeutic strategies for cancer. *Am. J. Pathol*. 170:793–804, 2007.

In: Advances in Zoology Research. Volume 6
Editor: Owen P. Jenkins
ISBN: 978-1-61122-061-2

Chapter 2

In the Wings

Marcia Brennan, Ph.D.
History and Religious Studies, Rice University, Houston, Texas
and Artist In Residence, Department of Palliative Care and Rehabilitation Medicine,
M. D. Anderson Cancer Center, Houston, Texas, US

Abstract

As a professor of the humanities at Rice University, and as an Artist In Residence in the Department of Palliative Care and Rehabilitation Medicine at the University of Texas M. D. Anderson Cancer Center, I have had the opportunity to witness the many dynamic qualities of butterflies. These remarkable beings appear at once as figures of strength and delicacy, freedom and interconnectedness, endurance and fragility. Inhabiting the terrestrial and the aerial domains simultaneously, butterflies are closely interwoven with their natural and social environments, just as they remain distinctive presences within them.

In this commentary, I present several vivid case studies in which butterflies appear in artworks produced by individuals at the very end of life. These moving words and images exemplify the ways in which butterflies embody key themes within aesthetics, spirituality, and the medical humanities. Throughout the stories butterflies appear as figures of transience and endurance that cross multiple boundaries within and beyond the end of life.

These materials thus offer an innovative perspective that complements the views of the scientific community, while contributing to a larger case for the environmental importance and cultural value of the interconnected lives of human beings and butterflies.

Introduction: "You Have to be Gentle with Everything That You Touch"

One afternoon I met a young woman who was in the final stages of liver cancer. She told me that her mother, who had maintained a nearly constant vigil by her bedside, was having an extremely hard time accepting her daughter's imminent death, so she was continually "busy trying to be busy, trying to be a mother. Because she's scared, she's become ADHD. So I told her, 'Just give me a kiss. Just stay connected.'" This gifted young woman was a painter, a textile artist, and a vocal performer. During our visit she drew a butterfly with her own hands, both literally and symbolically. That is, she used the network of linear indentations threading through the palm of her left hand as a guide to model the patterns appearing on the butterfly's wings. She then made a miniature portrait of herself holding onto a slender thread connecting her diminutive, dangling figure to the butterfly's skyward flight. Looking first at the drawing and then back at me, she emphasized, "You have to be gentle with everything that you touch."

Figure 1. Lyn Smallwood, *The Gossamer Thread*, 2012, pencil on Arches paper, 12 1/8 x 7 ½ in.

Reflecting the themes of this story, the visual artist Lyn Smallwood created an image of a butterfly with outstretched, patterned wings and delicate front legs that clasp the end of a gossamer thread (Figure 1). Floating gracefully on the otherwise blank page, the butterfly inhabits a middle ground that connects the realms of above and below, thereby uniting life on the ground and in the air. This suggestive pencil drawing thus visualizes the fragile links between multiple states of being. That is, while we cannot see what lies on the other end of the slender thread that dangles somewhere off the bottom of the page, we also cannot see what lies directly ahead, as the butterfly is poised to fly off in to an open space that we can only just imagine.

Artist In Residence

In my "day job," I am an Associate Professor of Art History and Religious Studies at Rice University, where my research areas include modern and contemporary art history and museum studies, comparative mysticism and metaphysics, and the emerging discipline of the medical humanities. In addition to my academic life at Rice, since early 2009 it has also been my privilege to serve as an Artist In Residence in the Department of Palliative Care and Rehabilitation Medicine at the University of Texas M. D. Anderson Cancer Center. In the hospital context, I work as a creative writer with individuals at the very end of life. Despite the differences between my research fields, each of these areas shares a common set of themes and challenges, namely: How do we find language to describe states of being for which there is no language? How do we represent the unrepresentable and translate the untranslatable? How do we formulate metaphors of transience to describe subjects that are simultaneously coming into form and going out of form, sometimes within the framework of a single image? And how do we represent the fragile zones of contingency that mark both gradual transitions as well as more dramatic shifts between states of being? Notably, these issues are as pertinent to theoretical discussions of abstract painting's simultaneously dissolving and crystallizing structures as they are to the contemplation of spiritual experience and mystical ecstasy, as they are to the very real challenges of people at the end of life. Thus, throughout all of my work I am fascinated—and profoundly moved—by subjects and situations that repeatedly exceed our capacity to represent them, even as we repeatedly attempt to do so, often through imagery that conveys transitional states and transformational visions.

At M. D. Anderson, my work in the Artist In Residence program is sponsored by COLLAGE: The Art for Cancer Network (http://www.collageartforcancer.org/home.htm), a nonprofit organization conceived and founded by Dr. Jennifer Wheler, Assistant Professor in the Department of Investigational Cancer Therapeutics. Working on both the Acute Palliative Care Inpatient Unit and on the Palliative Care Outpatient Clinic, I assist patients and their caregivers in constructing meaningful narratives to express their insights and experiences. Thus I primarily serve as a translator, yet much of what I translate is ineffable. I create a context, and provide critical skills, so that people facing extraordinarily difficult life situations have an opportunity for personal and symbolic expression, which then becomes clothed in aesthetic form.

As an Artist In Residence, I constantly confront the challenge of creating human connections across a formidable gulf of separation. Frequently, when I initially approach a patient I am told that the work "sounds interesting, but I have no artistic ability." This invites me to respond, "Well, let's just say for a moment that you *did* have artistic ability. What images would you want to write about?" Depending on the circumstances I might also ask, "Where are you from? What do you love to do?", or even more simply (yet never simply), "What do you love?" Sometimes I'll just ask people directly, "If you had an image in your mind of something that holds special meaning for you—and it can be anything at all—what would that be?" Very often a flash of illumination will become visible on the person's face, and they will share an image with me. They will describe a scene or a subject, while I gently encourage them to talk and make notes to help crystallize their thoughts. Once the artwork is complete, I read the person's words back to them, while making any additions or corrections that they indicate. The narrative is then inscribed into a handmade paper journal, which the person is able to keep and share with their family, either as a medium for further creative expression or as a legacy gift that performs a memorial function. The portable microenvironment of the text thus provides a durable yet tender memory of this transient moment in their lives. These themes represent the subject of my current book project, *Words Beyond Words: Finding Language at the End of Life* (forthcoming from Intellect Books, U.K., in association with the University of Chicago Press).

AT THE EDGE OF THE WINDOWPANE: EVANESCENCE, FREEDOM AND JOY

One afternoon I met a woman who was scheduled to begin a clinical drug trial the following morning. For her, butterflies represented an ultimate state of freedom, one that entailed the capacity to overcome obstacles and live in the sweetness of the moment. As she told me:

> Butterflies are all about freedom.
> For me, they represent the ability to see that,
> No matter what the problem is,
> It doesn't hold you back
> But you get out there and continue to move.
> I see this with butterflies.
> The multicolored butterflies are unique,
> And the ones with the big eyes are so haunting.
>
> Butterflies became important for me recently
> Along with the idea that,
> Butterflies are only here for such a short time,
> But they enjoy life so much.

In many ways, end of life experience entails the conjunction of multiple perspectives. In palliative care people will often draw inward, into a private interior domain, as they prepare to move out of life itself. Yet just as people pull inward, they also remain partially in the world. Thus as people pull in *and* look out onto a familiar reality, reality itself can take on an unfamiliar appearance, and often, a visionary quality. Moreover, just as people at the end of life are poised on a threshold that conjoins multiple realities, as an Artist In Residence I repeatedly stand on the thresholds of patients' rooms, asking if I can come in and visit with them. As I introduce myself to people I've never met before, I perch symbolically on the edge of a windowpane that both separates and connects multiple visions of reality.

In turn, during our visits I draw people out as they turn inward *and* as they look outward onto the world before them. Incorporating these multiple perspectives, the artworks represent a common meeting place that potentially integrates otherwise disparate domains while giving voice to people's experiences. In so doing, the artworks often present a composite perspective that simultaneously encompasses the presence and the absence of subjects that lie above and below various thresholds of recognition, including subjects that can be seen as too lowly and insignificant, or as too magnificent and profound, to be fully integrated within the familiar parameters of everyday life. I have repeatedly witnessed the instrumental capacity of aesthetics to serve as just such a dynamic edge where seemingly ordinary experience becomes expressed through transformational states of consciousness. While often conveyed in simple and direct language, the artworks can offer a complex glimpse of life at the edge of the windowpane.

As Many Butterflies As I Can

In this context butterflies often represent a lightness of being, a hint of floating color, and a sense of the continuation of life through all cycles of life. One day on the clinic I met a woman who told me about her cultivation of native plants that supported indigenous butterfly gardens. As she said, "it wasn't until later in life that I finally became aware of the gardener's passion," particularly for plants that would attract black swallowtail butterflies. She also expressed her hope that her garden would represent a living link for her family after she was gone. Her story is entitled:

As Many Butterflies As I Can

I like indigenous wildlife,
Especially butterflies, dragonflies, birds, and mammals.
I have passion flower vines growing in my yard.

My interest now is in planning gardens for butterflies.
There's a place for the native in home gardening.
It's something that's become a part of me.
I want to give myself to the gardening,
And in my given area, I want to attract
As many butterflies as I can.

While both grounded and in flight, butterflies represent consummate figures of transition and continuation, just as they reflect the delicate and powerful ability to occupy multiple domains at once.

Monarchs, Mothers, and Messengers

Throughout the artworks, butterflies symbolically embody the capacity for perceiving the natural world as a living symbol of grace. One day I met a middle-aged woman who had been admitted to the palliative care ward the previous day.

Notably, this woman was suffering from the same type of deadly cancer that had claimed her mother's life several years earlier. As we sat together, this woman began to tell me about her mother and the circumstances of her passing. Clearly, this woman still missed her mother and identified closely with her presence. As she said:

> My image is of butterflies, and my Mom.
> When my Mom was dying I begged her to come back and see me,
> And she said, "I don't want to scare you,
> But if I can, I'll come back as a monarch butterfly."
> So every time I see one, I know it's my Mom.
> My image is of a beautiful monarch butterfly
> Landing on a stargazer lily.
> That's my favorite flower.

As in life itself, at the end of life people frequently experience a strong desire for close personal connections, including ties to family members who have passed, and the need for a sense of continuity between worlds. In these tender circumstances, butterflies can appear as vibrant threads interweaving perennial concerns, just as they serve as symbols of presence and memory, faith and spirit, life and the afterlife.

Like A Butterfly, for His Glory

One day I met an older couple in which the wife had just been admitted to the Palliative Care Inpatient Unit. While this woman had devoted her life to being a wife and mother, her husband had enjoyed an extremely successful professional career. Even now, while clothed modestly in a hospital gown, this woman retained an elegant appearance and she still wore some of her diamond jewelry.

When I asked her what she loved, she quickly responded, "I love God and Jesus above everything else. I taught Bible School for some twenty-odd years. I found God as a young child and I was always devout, but the cancer has made me even more so." As our conversation unfolded I learned that, in addition to her love for her church and family, this woman also loved butterflies. Indeed, these fragile creatures were sacred to her and played an important role at transitional moments in her life.

Her story is entitled:

Like A Butterfly, For His Glory

I love God above everything,
And I love my family,
And anything having to do with nature.

Since I was diagnosed with cancer,
I've had a lot of experiences with butterflies.
My daughter was with me,
And she had to go back home.
It was early December—and pretty cool outside—
And there weren't many butterflies around.
But one day she called me on the phone and said,
"Mom, I'm sitting here, feeling really sad,
And my friend who was sitting there beside me said, 'Look down!'"
And there was a gorgeous yellow butterfly
Sitting on her salad.
It flew away,
And then—that one, or another—came back and landed.
I knew that God was talking to us.
He can take something ugly,
And make it beautiful.
My daughter knew the first butterfly was for me,
And the second one was for her.

When we were coming home after the first chemo trip,
I stopped the car and got out.
No sooner had I taken a few steps,
But a huge yellow swallowtail encircled us,
And it was saying, "Welcome Home!"
God was encircling us,
And it was just wonderful.

Another time, I was just sitting there,
Feeling depressed.
I couldn't move my legs.
And all of a sudden,
A gorgeous, light orange butterfly came up.
She landed there,
And she stayed for at least half an hour.
Every time I've needed a lift,
Or needed God's presence,
There's a butterfly there.

If God can take this ugly, fuzzy old thing
And turn it into something beautiful—
Like a butterfly—
Then God can take this ugly cancer

And make something beautiful
For His Glory.

He can, and He does.
The butterflies are like a sunset,
All for His Glory.

As the miniature becomes monumental in palliative care, the artworks reveal just how close we are able to get, and how much we are able to see, of presences that simultaneously evoke the continuity *and* the transformation of everyday lived perspective. Nowhere is this more apparent than in images of butterflies. Again and again, butterflies represent a sense of interconnectedness between the natural, social, and divine worlds. In turn, the artworks reveal a person's ongoing capacity for self-creation as they reflect on the joys of a living environment in which microcosms speak to macrocosms, and the ordinary and the numinous translate freely between one another.

Conclusion: Butterflies on the Doorframe

When someone passes away on the Palliative Care Inpatient Unit at M.D. Anderson, the nurse places a colorful butterfly on their doorframe, while a tiny butterfly appears next to the person's name on the overhead board near the nurses' station. The butterfly thus signals people to proceed with great sensitivity and caution and, in the words of the young woman whose story appears above, to "be gentle with everything that you touch."

I once asked one of the nurses about the origin of this symbol. He told me that the butterfly is unique to this particular ward and it is not a universal symbol within palliative care. The nurse also told me that the idea first came to him during a medical conference at which another nurse mentioned butterflies in her presentation. The image stayed in his mind, particularly because the butterfly garden on the M.D. Anderson campus is such a popular spot, and people tend to have such positive associations with butterflies. Then, one day on the ward, someone passed away exactly at shift change, before the oncoming staff members could be notified.

Not knowing what had just occurred, cleaning people came in to empty the trash, while food service delivered the lunch, and both activities disrupted the grieving family. The nurse recognized that a sign was needed alerting people to these delicate circumstances and the butterfly symbol was adopted.

With great tenderness, the nurse also told me that he will ask family members whether they wish to take the butterfly home with them, and they almost always say yes. One family even requested several butterflies, one for each of the many grandchildren. Indeed, in the fragile setting of acute palliative care, the butterfly is an apt symbol of something that you can hold onto, but not quite. The butterfly embodies a lightness of being that evokes both the life that is here with us now, and the one that is waiting, just in the wings.

PUBLICATION NOTE

This project has been favorably reviewed by two independent Institutional Review Boards (IRBs) at Rice University and at the M. D. Anderson Cancer Center. Notably, the subjects of this transdisciplinary project are universal, human, and deeply important; as such, they raise a host of compelling issues regarding subjectivity, sensitivity, visibility, representation, and compassion that must be processed within our culture. At the same time, due to privacy concerns, the narratives presented here are abstracted sketches that are intended to be generic and anonymous. Particular details relating to specific individuals have been omitted or altered, thereby assuring HIPPA compliance and preserving issues of confidentiality, particularly as specified under "The Privacy Rule," The Belmont Report, and the Department of Health and Human Services Office for Human Research Protections, including The Common Rule and subparts B, C, and D of the Health and Human Service specifications as outlined in the Code of Federal Regulations (CFR) at 45 CFR 164 and 165, which specifies the "safe harbor" method of de-identification.

In: Advances in Zoology Research. Volume 6
Editor: Owen P. Jenkins
ISBN: 978-1-61122-061-2

Chapter 3

COMPARISON BETWEEN BIOLOGICAL CONTROL WITH FUNGI (*DUDDINGTONIA FLAGRANS* AND *MONACROSPORIUM THAUMASIUM*) AND CHEMICAL CONTROL (MOXIDECTIN) OF GASTROINTESTINAL NEMATODIASIS IN BEEF CATTLE OF TROPICAL SOUTHEASTERN BRAZIL

R. C. L. Assis**, *F. D. Luns, J. V. Araújo and F. R. Braga
Federal University of Viçosa, Department of Veterinary Medicine, Laboratory of Parasitology, Viçosa Campus, Viçosa, Minas Gerais, Brazil

ABSTRACT

Sodium alginate pellets of two different nematode predatory fungi *Duddingtonia flagrans* and *Monacrosporium thaumasium* (biological control) and moxidectin (chemical control) were evaluated in the control of gastrointestinal nematodiasis in bulls. Four groups of ten 6-month-old male Nellore calves were kept in paddocks of *Brachiaria decumbens* for 12 months. Each animal of group A received 1 g/10 kg body weight (b.w.) of pellets of *D. flagrans* (0.2 g fungus/10 kg b.w.) twice per week, for 12 months. In group B, each animal received 1 g/10 kg b. w. of *M. thaumasium* pellets, twice per week, for 12 months. In group C, each animal received 1 ml/50 kg b.w. of the anthelmintic moxidectin (1%) in May, July and September 2010. Animals of the control group (group D) received no treatment. In the analysis of fecal egg count (eggs per g feces; EPG), groups A and B had a 56.67% and 47.8% smaller count, respectively, than that of the control group ($p<0.05$). Moxidectin reduced EPG by 93.8% in the first 6 months of the study (period of application) and by 12.4% in the last 6 months, indicating reinfection by nematodes. Biological control maintained a low EPG in the long term while chemical control resulted in a short-term effect. The percentage reduction of

* Corresponding author: R. C. L. Assis, P.H. Rolphs Avenue, Viçosa Campus, Department of Veterinary Medicine, Laboratory of Parasitology, Vicosa, Minas Gerais, Brazil. Zip code: 36570-000, telephone: +55 31 3899-1458, fax: +55 31 3899-1457, e-mail: rafaela.carolina@ufv.br.

infective third-stage larvae (L_3) in relation to the control group at distances of up to 20 and 20–40 cm from the fecal pat was 64.5% and 73%, respectively, for group A; group B showed percentage reductions of 47.3% and 58% at the same distances. *Duddingtonia flagrans* showed a larger percentage reduction of L_3 in the pasture compared with *M. thaumasium*. Moxidectin showed no reduction of larvae in the pasture. Treatment of bulls with nematophagous fungus and moxidectin integrated in pellets can be used as an alternative treatment of bovine gastrointestinal nematodiasis; however, biological control was more efficient than chemical control over the long term in the environmental conditions of the present study.

Keywords: Cattle Nematoda, control methods of Nematoda, *Duddingtonia flagrans*, *Monacrosporium thaumasium*, moxidectin, helminths, beef cattle

INTRODUCTION

Gastrointestinal parasitism by nematodes causes diseases and can kill ruminants throughout the world, but the greatest economic impact is on the reduction in growth of young animals, resulting in low productivity (Araújo et al., 2007). The indiscriminate use of anthelmintics has provoked the selection of helminth populations that are resistant to the different chemical groups used in cattle treatments (Mota et al., 2003). Besides, the chemical residues of such products have a negative impact on the environment and public health. The growing concern about food and environmental safety has required new strategies in the control of gastrointestinal nematodes in order to reduce the use of synthetic molecules in conventional farming and extensive organic livestock farming.

One of the possible alternatives that have been researched and proposed for use alone or integrated is biological control with predatory fungi of nematodes. Research on the nematode-trapping fungi *Duddingtonia flagrans* (Larsen et al., 1995; Sarkunas et al., 2000; Dias et al., 2007) and *Monacrosporium thaumasium* (Alves et al., 2003; Araújo et al., 2004) with cattle has demonstrated the potential of the organisms as biological control agents against the free-living stages of parasitic nematodes in livestock under experimental and natural conditions. However, most studies of biological control in cattle have been conducted in temperate regions (Larsen et al., 1995; Sarkunas et al., 2000) and with dairy cattle, which are usually placed in the pasture only in the grazing season. There are no previous studies evaluating biological control with *D. flagrans* or *M. thaumasium* in extensive systems of beef cattle and, due to weather conditions, maintained on pasture throughout the year.

Furthermore, there are no studies that have compared the effects of chemical and biological control in reducing nematodes of cattle, allowing the development of an integrated methodology of application of chemical and biological control. Worldwide, there have been several reports on bovine anthelmintic resistance to ivermectin and albendazole, and to the association of ivermectin and albendazole, levamisole (Cleale, 2004; Waghorn et al., 2006). The avermectins (ivermectin and doramectin) and milbemycins (moxidectin) are potent anthelmintic compounds known as endectocide molecules due to their activity against ecto- and endoparasites (Shoop et al., 1995). They have low mammalian toxicity and their formulations are convenient to use. Hence, they are extensively used worldwide in veterinary medicine.

Thus, two isolates of nematophagous fungi native to Brazil and the anthelmintic moxidectin were selected for this study. This is the first work in its field to evaluate the effectiveness of the nematophagous fungi *D. flagrans* and *M. thaumasium* by testing pellet formulations of these fungi in a sodium alginate matrix in the biological control of gastrointestinal nematodiasis in beef cattle reared in an extensive system, comparing their effectiveness with a traditionally used chemical control (moxidectin), under the climatic conditions of tropical southeastern Brazilian.

Materials and Methods

Study Area

The experiment was carried out on the experimental farm of the Federal University of Viçosa, located in the municipality of Florestal, state of Minas Gerais, southeast region of Brazil (19°53′22″ south latitude and 44°25′57″ west longitude), from May 2010 to June 2011. The paddock's topography is flat to hilly (29% flat, 54% undulating and 17% hilly), with an average altitude of 750 m and native vegetation of the forest–cerrado transition zone. The climate is tropical with a dry season (Köppen–Geiger climate classification: Aw), with an annual average maximum temperature of 28°C and an average minimum temperature of 13.9°C.

Fungal Cultures

Two isolates of the predatory fungi *D. flagrans* (AC001) and *M. thaumasium* (NF34A) were kept in test tubes containing 2% corn meal–agar (Difco®, USA) at 4°C in the dark. These isolates came from a Brazilian soil and belong to the mycology collection of the Federal University of Viçosa, Brazil. To induce the formation of fungal mycelium, culture discs of 5 mm in diameter were transferred to 250 ml Erlenmeyer flasks containing 150 ml of potato dextrose (Difco®, USA) liquid medium (pH 6.5) under agitation at 120 revolutions per min, in the dark, at 26°C, for 10 days. After this period, the mycelia were harvested with a platinum loop, and weighed in an analytic scale for the future production of the pellets, which were made in a sodium alginate matrix according to Walker and Connick (1983) and as modified by Lackey et al. (1993).

Treatment of Cattle

A total of 40 male Nellore calves, 6 months old, were previously dewormed with 7.5 ml albendazole/10 kg body weight (b.w.). After 15 days of anthelmintic treatment, the cattle were divided into four comparable groups (A, B, C and D) of 10 animals each. The bulls were weighed and allocated to four 15.0 ha paddocks of *Brachiaria decumbens*, naturally infested with gastrointestinal parasite helminths, due to previous grazing by young and adult animals. Each group was allocated to only one paddock without rotational grazing between the groups during the experimental period. In group A, each animal received 1 g of pellets containing the *D. flagrans* fungus/10 kg b.w., twice per week for a period of 1 year. In group B each animal received 1 g of pellets containing the fungus *M. thaumasium*/10 kg b.w., twice per week for a period of 1 year. In group C, animals were subcutaneously injected with moxidectin (1%), according to the manufacturer's instructions (0.2 mg/kg; Cydectin, Fort Dodge®), in May, July and September 2010. Group D animals, the control group, received no treatment.

Processing of Fecal Samples

After the introduction of the bulls into the paddocks, samples of feces were collected, once per week, directly from the rectum to determine the number of nematode eggs per g feces (EPG), according to Gordon and Whitlock (1939). Fecal samples were collected from the animals to observe the growth of the fungi, once per week, 2 days after treatment with fungi. The feces were incubated in plates containing 2% water–agar and 100 infective third-stage larvae (L_3) recovered from the coprocultures and put into a drying oven at 25°C for 10 days to confirm the passage and the predatory ability of the fungi through the gastrointestinal tract of cattle and growth in the feces. At the same time as the EPG examination, coprocultures were carried out for each animal, according to methodology described by Roberts et al. (1952). The identification of the L_3 in the coprocultures was performed according to Keith (1953). EPG and L_3 recovered from coprocultures of animals from both treated and control groups were recorded, and the percentage of larval reduction was determined according to Mendoza-De-Guives et al. (1999): reduction (%) = mean L_3 recovered from control group − mean L_3 recovered from treated group × 100/mean L_3 recovered from the control group.

Pasture Samples

Every 15 days, herbage samples were collected in the paddock of each group, in a zigzag pattern from alternated points, at distances of 20 cm and 20–40 cm from fecal pats, according to Amarante et al. (1996). Then, a 500 g herbage sample was weighed, and parasitic nematode larvae were recovered following the procedure of Lima (1989). The samples were incubated in a drying oven at 100°C for 3 days to determine the dry matter content. Data were transformed into larvae per kg dry matter.

Treatment of Data

Meteorological data were obtained at the specialized station located in the Florestal region. The monthly means of maximum, medium and minimum temperatures, relative humidity and rainfall are reported. EPG, the number of L_3 recovered from the feces and paddocks, and correlation between EPG and recovered L_3 were compared over the experimental period. The weight of the animals was compared during the months of the experiment, starting from June 2010. EPG data were transformed into log (x + 1) prior to analysis. Data were examined by analyses of variance and Tukey's multiple-comparison test with 1% probability.

RESULTS AND DISCUSSION

The monthly average values of EPG counts are shown in Figure 1. In the first 3 months of the experiment (May, June and July 2010) there was no significant difference (p>0.05) between the groups treated with fungi (A and B) and the control group (D). From the month of August, group D (control) presented higher averages for EPG than groups A and B, until the end of the experiment (June 2011). In the last 6 months (January–June 2011) a statistical difference was observed (p<0.05) between the groups treated with fungi (A and B) and the moxidectin-treated and control groups (C and D). At the end of the experiment, mean monthly EPG from group A was 56.7% lower than the EPG of the animals of group D (control). In a study with the same *D. flagrans* isolate (AC001), Dias et al. (2007) obtained a 31% reduction in EPG of treated crossbred Holstein–Zebu cattle. In group B, treated with *M. thaumasium*, the reduction in mean EPG compared with the control group was 47.8% (p<0.05). Alves et al. (2003), in a study with crossbred Holstein–Zebu cattle treated with the same isolate of *M. thaumasium* obtained an 88.8% reduction in EPG after 4 months of their trial. There was a statistically significant difference (p<0.05) between the fungi *D. flagrans* and *M. thaumasium* on August 2010 and February, March, April, May and June 2011. EPG decreased significantly more in *D. flagrans* than in *M. thaumasium* at the end of the experiment.

Group C presented a particular EPG curve during the experiment. In group C, a reduction in EPG count was rapidly observed after the treatment with moxidectin. Unlike groups A and B, which showed a gradual reduction in the elimination of eggs, animals treated with moxidectin showed variations in the curve of EPG over the course of the study. Moxidectin prevents reinfestation by nematodes for a minimum period of 28 days, but does not prevent reinfestation over a longer period of time (Cleale, 2004). Thus, in the first 6 months of the experiment when the animals received doses of moxidectin at monthly intervals (May, July and September 2010), the reduction in mean EPG of group C compared with control was 93.8% (p<0.05), but within 6 months of the study EPG reduction was 12.4%, since during these months the animals received no new applications of anthelmintics and were reinfected by nematodes. According to Cleale (2004) the means of fecal EPG counts from cattle treated with injectable moxidectin were reduced by 99.8% at 14 days after treatment, by 99.1% at 28 days after treatment and by 96.7% at 55/56 days after treatment. Soutello et al. (2010), in an experiment with the same drug (Cydectin®, Fort Dodge), obtained a reduction in EPG of 98.3% at 36 hours post-treatment. However, in these studies, the authors did not evaluate the reduction of EPG for a long period after treatment with moxidectin.

The percentage compositions of L_3 of different nematodes obtained after larval culture are presented in Table 1. The prevalence of *Cooperia* sp. was greater than that for the other nematodes in all groups, with percentages of 48.4, 53.1, 69.5 and 44.1% for the groups A, B, C and control, respectively, followed in descending order by *Haemonchus* sp. and *Oesophagostomum* sp. *Haemonchus* sp. were 38.2, 34.7, 17.4 and 43.4% and *Oesophagostomum* sp. were 12.1, 10.9, 11.5 and 10.4%, in the four groups, respectively.

Genera found in smaller quantities and sporadically in the four groups were *Bunostomum* sp. (0.8, 0.6, 1.0 and 1.2%) and *Strongyloides* sp. (0.5, 0.7, 0.6 and 0.9%). The results may indicate the presence of moxidectin resistance after treatment (group C). The genus *Cooperia* probably showed resistance to moxidectin, based on larval identification. A difference was observed in the percentages of *Cooperia* sp. in relation to the genera *Haemonchus* and *Oesophagostomum* of the animals in group C (Table 1). During the period when the animals were receiving applications of moxidectin the percentage of *Cooperia* was significantly higher than for other genera in this group.

Another finding indicating resistance of *Cooperia* to moxidectin is presented in the comparison of recovery of this genus relative to other groups in the months from May to October 2010. Although the genus *Cooperia* has been prevalent in the study, in these months there was a distinct emphasis on the percentage of these larvae compared with the percentage found in all the other groups (A, B and D). In group C (moxidectin) *Cooperia* showed no progressive reduction in its recovery, as in the groups treated with fungi (A and B), as the fungi are not selective between nematode larvae and reduced the number of larvae of all genera.

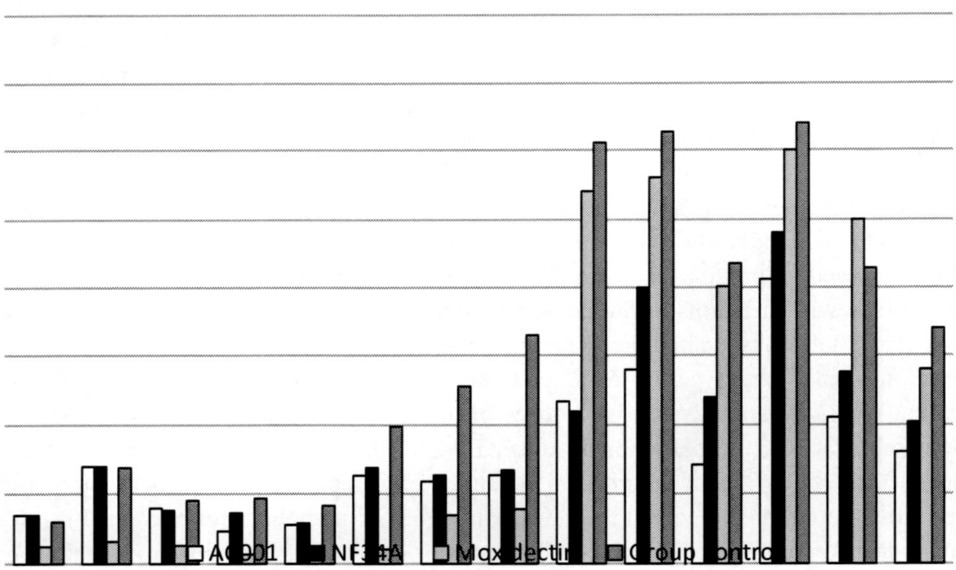

Figure 1. Monthly average counting of eggs per g feces (EPG) for the groups treated with the nematophagous fungi *Duddingtonia flagrans* (AC001) and *Monacrosporium thaumasium* (NF34A) (1 g fungus/10 kg body weight), for the group treated with moxidectin (May, July and September 2010) and for the control group, collected from May 2010 to June 2011 in Florestal, Minas Gerais, Brazil.

Currently, anthelmintic resistance has become a threat to the prophylaxis of parasitic gastroenteritis in cattle worldwide (Soutello et al., 2007). In Brazil, there are several reports on bovine anthelmintic resistance. A high frequency of ivermectin-resistant *Haemonchus* spp. and *Cooperia* spp. has been detected and in some cases albendazole-resistant nematodes were also found (Soutello et al., 2007; Souza et al., 2008). According to Codin et al. (2009), moxidectin presented 100% efficacy against the genera *Haemonchus* and *Trichostrongylus*, but 65.2% efficacy against *Cooperia* spp. Soutello et al. (2007), in a study with cattle in Brazil, related that moxidectin-resistant *Cooperia* species were detected (*Cooperia punctata* and *C. pectinata*). *Cooperia* species are involved in most of the reports concerning anthelmintic resistance in cattle. Reduced moxidectin efficacy against larval stages of *C. punctata* (Williams and DeRosa, 2003) and against adult stages of *C. punctata* (88.4%), *C. spatulata* (93.2%), *C. oncophora* (88.7%) and *C. surnabada* (89.1%) (Ranjan and Delay, 2004) was reported in cattle in the USA. These studies indicate that the capacity of *Cooperia* to survive or develop resistance to the macrocyclic lactone anthelmintics is apparently higher than that of the remaining nematode species. Comparing the fungi *D. flagrans* (group A) and *M. thaumasium* (group B) there was no difference between the nematode genera found. These isolates showed no selectivity for a particular genus.

Table 1. Percentage values corresponding to infective larvae (L_3) recovered from the coprocultures of the groups treated with the nematophagous fungi *Duddingtonia flagrans* (group A) and *Monacrosporium thaumasium* (group B) (1 g fungus/10 kg body weight), treated with moxidectin (May, July and September 2010) (group C) and of the control group (group D), collected from May 2010 to June 2011 in Florestal, Minas Gerais, Brazil. (Coop. = *Cooperia*, Haem. = *Haemonchus*, Oeso. = *Oesophagostomum*)

	Group A			Group B			Group C			Group D		
	Coop	Haem	Oeso	Coop	Haem	Oeso	Coop	Haem	Oeso	Coop	Haem	Oeso
May	45	55	0	39	61	0	87	4	9	61	39	0
Jun	29	71	0	21	71	4	84	8	8	75	25	0
Jul	62	35	3	72	24	0	75	10	15	41	53	6
Aug	68	27	5	74	18	3	78	14	8	29	63	8
Sep	75	11	14	82	10	2	69	11	20	22	68	10
Out	65	25	10	75	8	17	59	24	13	22	60	18
Nov	76	11	9	77	9	14	55	25	14	17	56	7
Dec	36	8	56	61	29	8	63	27	10	18	52	25
Jan	55	33	12	41	10	49	49	34	15	40	41	19
Feb	44	25	31	61	25	14	77	10	13	35	29	32
Mar	50	34	6	76	4	20	71	8	21	46	49	5
Apr	32	54	14	14	77	9	49	37	8	69	24	7
May	26	64	10	28	67	5	76	19	5	63	28	9
Jun	15	82	0	23	70	7	81	12	2	79	21	0
Average	48.4	38.2	12.1	53.1	34.5	10.9	69.5	17.4	11.5	44.1	43.4	10.4

The larval numbers of *Cooperia*, *Haemonchus* and *Oesophagostomum* recovered from the pasture (per kg dry matter) of the group treated with *D. flagrans* reduced by 60.5, 53.2 and 47.5%. In the group treated with *M. thaumasium* the reduction was 33.6, 26.6 and 57.5%, respectively ($p<0.01$). *Monacrosporium thaumasium* provided a lower percentage reduction in the pasture compared with *D. flagrans*. The group treated with moxidectin showed no reduction in the number of larvae on pasture compared with the control group. Fontenot et al. (2003) reported that *D. flagrans*, in addition to reducing the number of L_3 in the pastures, could also account for the reduction of the L_3 burden in animals that would be introduced into the pasture in the future. Isolates of *Duddingtonia* reduced the level of L_3 in herbage by 74–85% (Grønvold et al., 1993). In all groups, the number of L_3 recovered up to 20 cm differed from that found 20 and 40 cm from the fecal pat. Of the total L_3 recovered from the paddocks, 84% were in the distance of up to 20 cm from the fecal pat and only 16% were recovered within 20–40 cm from the fecal pats in the group treated with *D. flagrans*. The same pattern was found in the group treated with *M. thaumasium*, where 91% of the larvae were up to 20 cm from the fecal pat and only 9% from 20 to 40 cm. Acording to Souza et al. (2008), in the dry season, most of the larvae remain in the stool, or near the base of grass. The opposite occurs during the rainy season when the larvae can migrate 90 cm around the stool. In the moxidectin and control group the highest percentage of L_3 was also up to 20 cm from the fecal pat (70 and 73%), but there was a greater amount of L_3 recovered from 20 to 40 cm (30 and 27%) compared with the treated groups. The percentage reduction of L_3 in relation to the control group in the distances of up to 20 and 20–40 cm from the fecal pat was 64.5% and 73%, respectively, for group A; group B showed percentage reductions of 47.3% and 58%, for the same distances. *Duddingtonia flagrans* showed a larger percentage reduction of L_3 in the pasture compared with *M. thaumasium* (p>0.01), but no difference was observed between groups A and B in October, November and December 2010 and January 2011. The group treated with moxidectin showed no statistical difference in the number of recovered larvae on pasture compared with the control group, and thus there was no effect in reducing or preventing reinfestation through pasture. This result was expected since moxidectin acts by reducing the parasitic load of the animals in the short term only and at this point it is appropriate to integrate with a biological control program.

Climatic data are presented in Figure 2. The monthly average temperature ranged from 15.9°C (July 2010) to 27.2°C (March 2011).

The optimum temperatures for the development of most gastrointestinal nematode parasites of cattle (Trichostrongylidae) are between 20 and 30°C (O'Connor, 2006). Average temperatures from 16 to 20ºC are suitable for the development of most parasitic nematode larvae of cattle in pastures. The optimum temperature for the growth and predatory activity of nematophagous fungi is between 20 and 30°C (Su et al., 2007). In the climatic conditions of this study, temperature had no effect in limiting larval development because the average temperatures observed favor the growth and action of the fungal isolate for almost all the experimental period. From May to July 2010 and from April to June 2011 (dry season) rainfall was below 50 mm. The minimum rainfall required for larval development in the environment is 50 mm (Boom and Sheath, 2008). During June 2010 and June 2011, temperature and rainfall were unfavorable for egg and larval development in pastures. Almost no larvae were recovered in the pastures during these months, but even when the rainfall was very low or zero, there was reinfection of the animals, since eggs were recovered in the feces. These results indicate that humidity in the stool allows the development of nematodes in their free-living stages and that low rainfall is sufficient for the development and larval migration to the pastures. Van Dijk et al. (2009) noted that infection levels are highest during the rainy season and related to higher humidity, which favors the development of parasites in their free-living stages and L_3 migration from the stool to adjacent pastures. The low larval occurrence on pasture in periods without precipitation and scores higher in the rainy season observed in this study are consistent with these observations. In months where temperature and humidity are not favorable for the development of fungus and nematode larvae on pasture, it would not be necessary to use the fungus. This study raises the possibility that in the case of consortium application of chemical and biological control, perhaps the use of biological control could be omitted in the months of the initial chemical applications, when the anthelmintic in question is able to reduce infective load by about 90%. However, taking into account the issue of the resistance of some genera to anthelmintics, the ideal would be to not omit the application of fungus. This study highlights the need for future assessments of biological control syndicated to chemical control, at different dosages and climatic conditions.

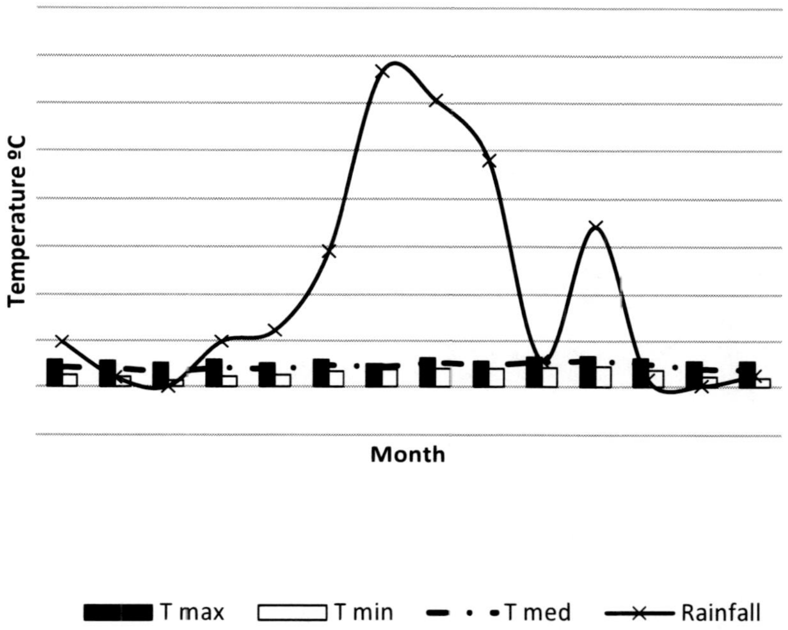

Figure 2. Monthly average maximum (T max), minimum (T min) and medium (T med) temperatures, and average monthly pluviometric gradient in Florestal, Minas Gerais, May 2010 to June 2011.

In relation to weight gain, all groups showed continued growth in weight during the rainy season and lost weight during the dry season, associated with a marked reduction in growth and nutritive value of the pasture (Figure 3). Even so, a difference was observed in the weight gain of animals of the treated groups in comparison with the animals of the control group. This reinforces the fact that the administration of pellets with fungi favored the health of the animals, with reductions of L_3 in the pasture where the animals of this group were likely to be contaminated by a lower L3 burden. A higher weight gain in the animals in relation to the control group was also observed by Araújo et al. (2007) when testing the fungus *M. thaumasium* in goats in a semi-arid region of Brazil. Braga et al. (2009) in a field study with horses using the nematophagous fungus *D. flagrans* observed significant differences in weight gain between the groups treated with the fungus and the control group. In the present study, in the period from March to June the rate of weight increase did not appear to be much different in the two fungus-treated groups. There was a significant difference (p>0.01) for weight gain between the two treated groups only in 3 months of the study (July, October and November 2010), when the animals treated with *D. flagrans* showed a higher mean weight.

The group treated with moxidectin showed greater weight gain than the groups treated with the fungus in the months of May, July and September 2010, which were the months in which calves received applications of the anthelmintic. According to the data of Cleale (2004), the rate of weight gain by cattle treated with long-acting injectable moxidectin was 0.59 kg/day or 23% (0.11 kg/day) more than controls.

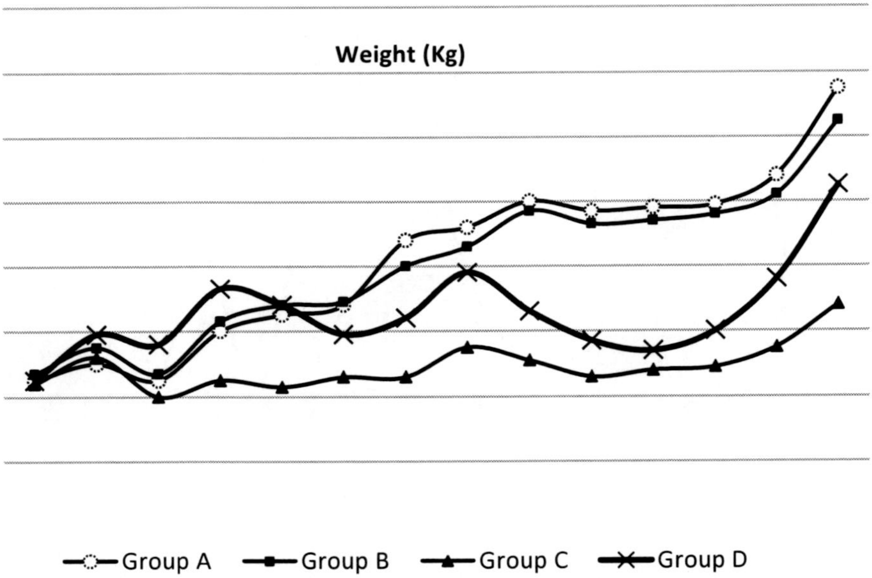

Figure 3. Mean weight gain (kg/day) of the groups treated with *Duddingtonia flagrans* (group A), *Monacrosporium thaumasium* (group B), moxidectin (group C) and of the control group (group D) from May 2010 to June 2011, Florestal, Minas Gerais, Brazil.

This study raises the possibility that in the case of consortium application of chemical and biological control, perhaps the use of biological control could be omitted in the months of the initial chemical applications, when the anthelmintic in question is able to reduce the infective load by about 90%. On the other hand, taking into account the issue of resistance among genera to anthelmintics, the ideal would be to not omit the application of fungus. This study brings the need for future assessments of biological control syndicated to chemical control, at different dosages and climatic conditions. According to Waller and Larsen (1993), the application of fungi in nematode biocontrol helps the chemical control and should be administered not only when there is a prediction of greater pasture infestation by helminth eggs and larvae, but also when there would be better conditions for the fungi growth in the environment, this way preventing the clinical parasitism and productivity losses, already supplying a sufficient number of larvae to allow these animals to develop a naturally acquired immunity. The results were analysed based on the overall experimental period and we conclude that this dose and periodicity of application of fungus pellets makes the use of other anthelmintic treatments unnecessary.

REFERENCES

Alves, P.H, Araújo, J.V, Guimarães, M.P, Assis, R.C.L, Sarti, P., Campos, A.K., 2003. Control of bovine gastrointestinal nematodes using formulation of the nematode-trapping fungus *Monacrosporium thaumasium* (Drechsler, 1937). *Arq. Bras. Med. Vet. Zootec.* 55, 568-573.

Amarante, A.F.T., Padovani, C.R., Barbosa, M.A., 1996. Contaminação de larvas de nematoídes gastrintestinais parasitos de bovinos e ovinos em Botucatu-SP. *Rev. Bras. Parasitol. Vet.* 5, 65-73.

Araújo, J.V., Guimarães, M.P., Campos, A.K., Sá, N.C., Sarti, P., Assis, R.C.L., 2004. Control of bovine gastrointestinal nematode parasites using pellets of nematode trapping fungus *Monacrosporium thaumasium*. *Ciênc. Rural* 34, 457-463.

Araújo, J.V., Rodrigues, M.L.A., Silva, W.W., Vieira, L.S., 2007. Controle biológico de nematoides gastrintestinais de caprinos em clima semi-árido pelo fungo *M. thaumasium*. *Pesqui. Agropecu. Bras.* 42, 1177-1181.

Boom, C.J., Sheath, G.W., 2008. Migration of gastrointestinal nematode larvae from cattle faecal pats onto grazable herbage. *Vet. Parasitol.* 157, 260–266.

Braga, F.R., Araújo, J.V., Silva, A.R., Araújo, J.M., Carvalho, R.O., Tavela, A.O., Campos, A.K., Carvalho, G.R., 2009. Biological control of horse cyathostomin (Nematoda: Cyathostominae) using the nematophagous fungus *Duddingtonia flagrans* in tropical southeastern Brazil. *Vet. Parasitol.* 163, 335-340.

Cleale, R.M., Hart, K.B., Hutchens, D.E., Johnson, E.G., Paul, A.J., Smith, L.L., 2004. Effects of subcutaneous injections of a long acting moxidectin formulation in grazing beef cattle on parasite fecal egg reduction and animal weight gain. *Vet. Parasitol.* 126, 325-338.

Condi, G.K., Soutello, R.G.V., Amarante, A.T., 2009. Moxidectin-resistant nematodes in cattle in Brazil. *Vet. Parasitol.* 161, 213–217.

Dias, A.S., Araújo, J.V., Campos, A.K., Braga, F.R., Fonseca, T.A., 2007. Application of a formulation of the nematophagous fungus *Duddingtonia flagrans* in the control of cattle gastrointestinal nematodiosis. *World J. Microbiol. Biotechnol.* 23, 1245-1252.

Fontenot, M.E., Miller, J.E., Peña, M.T., Larsen, M., Gillespie, A., 2003. Efficiency of feeding *Duddingtonia flagrans* chlamydospores to grazing ewes on reducing availability of parasitic nematode larvae on pasture. *Vet. Parasitol.* 118, 203-213.

Gordon, H.M., Whitlock, H.V., 1939. A new technique for counting nematode eggs in sheep faeces. *J. Counc. Sci. Ind. Res.* 12, 50-52.

Grønvold, J., Wolstrup, J., Nansen, P., Henriksen, S.A., Larsen, M., Bresciani, J., 1993. Biological control of nematode parasites in cattle with nematode-trapping fungi: a survey of Danish studies. *Vet. Parasitol.* 48, 311-325.

Keith, R.K., 1953. The differentiation on the infective larvae of some common nematode parasites of cattle. *Aust. J. Zool.* 1, 223-235.

Lackey, B.A., Muldoon, A.E., Jaffe, B.A., 1993. Alginate pellet formulation of *Hirsutella rossiliensis* for biological control of plant-parasitic nematodes. *Biol. Cont.* 3, 155-160.

Larsen, M., Nansen, P., Wolstrup, J., Grønvold, J., Henriksen, S.A., Zorna, A., 1995. Biological control of trichostrongyles in calves by the fungus *Duddingtonia flagrans* fed to animals under natural grazing conditions. *Vet. Parasitol.* 60, 321-330.

Mendoza-De-Guives, P., Davies, K.G., Clarck, S.J., Behnke, J.M., 1999. Predatory behaviour of trapping fungi against srf mutants of *Caenorhabditis elegans* and different plant and animal parasitic nematodes. *Parasitology* 119, 95-104.

Mota, M.A., Campos, A.K., Araújo, J.V., 2003. Controle biológico de helmintos parasitos de animais: estágio atual e pesrpectivas futuras. *Pesq. Vet. Bras.* 23, 93-100.

O'Connor, L.J., Walkden-Brown, S.W., Kahn, L.P., 2006. Ecology of the free-livingstages of major trichostrongylid parasites of sheep. *Vet. Parasitol.* 142, 1–15.

Ranjan, S., DeLay, R., 2004. Therapeutic and persistent efficacy of moxidectin 1% non aqueous injectable formulation against natural and experimentally induced lung and gastrointestinal nematodes in cattle. *Vet. Parasitol.* 120, 305–317.

Roberts, F.H.S., O'Sullivan, P.J., Riek, R.F., 1952. The epidemiology of parasitic gastro-enteritis of cattle. *Aust. J. Agr. Res.* 3, 187-226.

Sarkunas, M., Larsen, M., Nansen, P., Hansen, J.W., 2000. Biological control of trichostrongylid infections in calves on pasture in Lithuania using *Duddingtonia flagrans* a nematode-trapping fungus. *J. Helminthol.* 74, 355-359.

Shoop, W., Mrozik, H., Fisher, M., 1995. Structure and activity of avermectins and milbemycins in animal health. *Vet. Parasitol.* 59, 139–156.

Soutello, R.V.G, Seno, M.C.Z., Amarante, A.F.T., 2007. Anthelmintic resistance in cattle nematodes in northwestern São Paulo State, Brazil. V*et. Parasitol.* 148, 360-364.

Souza, A.P., Ramos, C.I., Bellato, V., Sartor, A.A., Schelbauer, C.A., 2008. Resistência de helmintos gastrintestinais de bovinos a anti-helmínticos no Planalto Catarinense. *Cienc. Rural* 38, 1363–1367.

Su, H., Hao, Y., Mo, M., Zhang, K., 2007. The ecology of nematode-trapping hyphomycetes in cattle dung from three plateau pastures. *Vet. Parasitol.* 144, 293–298.

Van Dijk, J., De Louw, M.D.E., Kalis, L.P.A., Morgan, E.R., 2009. Ultraviolet light increases mortality of nematode larvae and can explain patterns of larval availability at pasture. *Int. J. Parasitol.* 39, 1151–1156.

Waghorn, T.S., Leathwick, D.M., Rhodes, A.P., Jackson, R., Pomroy, W.E., West, D.M., Moffat, J.R., 2006. Prevalence of anthelmintic resistance on 62 beef cattle farms in the North Island of New Zealand. *N.Z. Vet. J.* 54, 278–282.

Walker, H.L., Connick, W.J., 1983. Sodium alginate for production and formulation of mycoherbicides. *Weed Sci.* 31, 333-338.

Waller, P.J., Larsen, M., Faedo, M., Henessy, D.R., 1994. The potential of nematophagous fungi to control the free-living stages of nematodes parasites of sheep: *in vitro* and *in vivo* studies. *Vet. Parasitol.* 51, 289–299.

Williams, J.C., DeRosa, A., 2003. Dose confirmation of moxidectin 0.5% pour-on against adults and fourth-stage larvae of various *Cooperia* spp. and *Trichostrongylus colubriformis* in Louisiana. *Vet. Parasitol.* 114, 295–303.

In: Advances in Zoology Research. Volume 6
Editor: Owen P. Jenkins
ISBN: 978-1-61122-061-2

Chapter 4

ADVANCES IN ZOOGEOGRAPHY: MICRO-SPACES OF AVIAN FORAGING AND HUMAN PRESENCE

Michael O'Neal Campbell
Lakehead University, Thunder Bay, Ontario, Canada

ABSTRACT

Traditional zoogeographical theory focuses on global regions, with less attention to small spaces shared by people and animals. Two relevant, possibly opposite, research strands, the statistical analysis of alert/flight distances and the subjective 'actancy' strand within animal geography are rarely examined as complementary to a new look at micro-spaces of human/avian relations within the larger theoretical framework of zoogeography. Alert distances (the distance between a bird and an approaching human when the bird shows awareness), flight distances (the distance, in such a scenario when the bird takes flight) and alert periods (the period between first alertness and flight) vary according to cover structure, elevation and proximity, human and road proximity. This chapter argues that while these parameters give a generalized contribution to the sharing of micro spaces, a strand of animal geography that emphasizes the unpredictable, subjective, individualist behavior of animals (birds in this study) offers a complementary perspective to the more classical statistical prediction of alert/flight distance based analyses. Illustrative examples are four papers published by this author; two on the alert distances of birds in the green spaces of Stirling, Scotland and Peterborough, Canada; two papers on the utility of a strand of animal geography (emphasizing avian individualism) in feeding spaces of Glasgow, Scotland and Peterborough, Canada; and one paper on the landscape ecology of birds in Ottawa, Canada. It is concluded that for effective environmental applications, the study of alert and flight distances must acknowledge bird (intra- and inter- species) adaptation and habituation to human presence; the lack of such acknowledgment may reduce the validity of distance analysis in avian studies, with implications for conservation and landscape planning. This represents an important contribution to the sub-discipline of zoogeography and to its parent disciplines of zoology and geography.

THE DEVELOPMENT OF ZOOGEOGRAPHICAL THEORY

Biogeography, the study of the distribution of species and ecosystems in geographic space and through geological time, is a "synthetic science", linking with ecology, evolutionary biology and palaeontology (McDonald 2003: 3). Its sub-disciplines also link to the environment (ecological biogeography), time (historical biogeography), explanation of species distribution and evolution (analytical biogeography) and the application of these strands to methodologies of protection (conservation biogeography). Biogeography is also divided in terms of life forms into zoogeography, concerned with the geographic distribution (present and past) of animal species; and phytogeography, concerned with the geographic distribution of plant species and their influence on the earth's surface. Zoogeography, as the interface of zoology and geography, shares methodological approaches and sub-disciplines of both of these sciences ((McDonald 2003; Campbell 2009a).

Zoogeography is historically justified; "knowledge of the geographical distribution of animals, and their faunal relations past and present, constitutes a very important part of the science of zoology" (Oilman et al. 1905). Initial definitions of zoogeography focused on regional or global dynamics. For example, Ingersoll (1920), defines the subject as "the science of mapping out the surface of the earth with reference to its faunæ; the study of the distribution of animal life", this concerned with "local peculiarities in animals and plants" in the "various regions of the earth's surface". Hence, species dispersal and introductions, obstacles to animal diffusion such as mountains and seas, changes in such barriers, climate and zoogeographical regions were important aspects of zoogeography (e.g., Sclater 1858; Wallace 1876; Drude 1884; Heilprin 1887; Beddard 1895). Nevertheless, zoogeography is comparatively under-studied; currently animals are a "paradox in biogeography"; they are popular, but there is a "relative scarcity of texts in animal biogeography" (Ganderton and Coker 2005: 123).

Also under-studied are methodological approaches that support micro-scale interfaces between zoogeography and human geography (Campbell 1998, 2009, 2013a,b,c; McDonald 2003). Nevertheless, two important recent developments have occurred, largely outside the discipline, offering the opportunity of integrating the global and regional analysis of traditional zoogeography and that of the micro-scale spaces heavily influenced by social structures and actors, and the shared coevolution of human behavior and animal adaptation.

One of these developments was the evolution of the 'New Ecology' (Holling 1973, 1995; Weins 1984; De Angelis and Waterhouse 1987; Stott 1994; Pickett et al. 1992; Zimmerer 1994; Sullivan 1996; Rowe 1997; Scoones 1997; Campbell 1998; 2013a,b,c). Increased dynamisim was injected into the earlier paradigm that was rooted in Newtonian physics, which favoured "deterministic dynamics" of nature, caused by a "few simple laws" (Solbrig 1994, 17), generalisation, scientific universalism, a final ecological equilibrium, mechanistic succession, vegetation climaxes and human/nature separation (Odum 1971; Pickett et al. 1992; Solbrig 1993; Begon et al. 1996).

This broader approach complements the older ecological paradigm, by expressing continual flux rather than stability and predictability in ecosystems. The emphasis shifts to unpredictable, non-linear, non-equilibrium and multi-directional spatial ecological change as inherent and subject to "multiple interpretations" (Nicolis 1994; Scoones 1997, 162; Hobbs et al. 2009). Environmental changes are evaluated as possibly not reversible or predictable

(Solbrig 1993; Levy-Tacher and Golicher 2004). As chaotic changes are intrinsic and internal (rather than externally derived) to the ecological system, complex change may be studied as an aspect of "non equilibrium or multi-state systems" under continual change (Young and Solbrig 1993: 324).

The prevalence of this strand developed over several decades (e.g., Holling 1973, 1995; Noy-Meyer 1975; Weins 1984; De Angelis and Waterhouse 1987; Westoby et al. 1989; Stott 1994; Pickett et al. 1992; Sullivan 1996; Rowe 1997; Scoones 1997; Campbell 2013a). Increasingly, ecologists and biogeographers have paid attention to the socio-environmental, multifaceted and continually changing roots of ecosystem health, which is defined by Karr and Dudley (1981: 21) as "the capability of supporting and maintaining a balanced, integrated, adaptive, community of organisms having species composition, diversity, and functional organization comparable to that of natural habitats of the region" (see also Clark 1973; Ludwig et al. 1993; De Leo and Levin 1997; Campbell 2013a).

Another recent (1960s) development concerned certain strands within 'animal geography', which modeled animals as subjects with active abilities (actancy, similar to people) rather than as mechanistic objects in human life spaces (Wolch and Emel 1998; Lorimer 2007; Campbell 2009a,b; Barua 2013). Here, actancy is defined as a human behavior, in which free choices are made in response to different events (Murdoch 1998). This strand, similar to the disequilibrium paradigm, argues for a more flexible perspective to uncover subjectivities undetected by the more positivist science of zoogeography, linking more effectively with the spatial human geographies of wildlife (Sauer 1952; Bennett 1960; Isaac 1970; Anderson 1997; Philo and Wilbert 2000; Yarwood and Evans 2000; Emel et al. 2002). The acknowledgement of subjective animal behavior allows more detailed comparisons between animal and human behavior (Wolch and Emel 1998).

Therefore, frequent characterizations of animal geography and zoogeography as synonymous neglect differences in some strands (Cansdale 1952; Bennett 1960; Campbell 2009a,b). Classical zoogeography is dominated by or even exclusively concerned with positivist quantification of animal presence, and also emphasizes large, regional or continental spaces (Campbell 2013a). The different approach of the dynamic strand of animal geography includes a greater theoretical focus on unique animal behavioral subjectivity and includes strands of 'cultural human geography', the hypothesis of shared actancy between people and animals and the integration of these with quantitative zoogeographical field methods (Wolch and Emel 1998; Johnston et al. 2000). The contribution of these to zoogeographical paradigms has been validated by some recent studies (Campbell 2008, 2009a, b).

The Zoogeography of Micro-Spaces

Debates on animal actancy and generalist behaviors are necessarily rooted in micro-spaces of intense human/animal contact; however a zoogeographical study bias towards large areas or regions exists (Campbell 2010). Micro-space studies are however developing within both zoogeography and human geography, because animals increasingly share small spaces with people (Hovorka 2008; Buller 2014; Hodgetts and Lorimer 2014). In the social sciences, such small scale studies are mostly termed actor-oriented, in contrast to macro-level, structural analysis. Actor oriented orientation are relevant to zoogeographical micro-studies, as at this refined scale, variable human behavior assumes prominence. Individual animal adaptation is also increasingly apparent (Wolch and Emel 1998; Urbanik 2012; Emel and Urbanik 2010). Studies must determine the relative importance of generalisable factors (species, size, environmental structures) for animal reactions to variable human behavior (e.g., noise, hand movements, approaches with and without focused intent) and the individual animal behavior that illustrates actancy (Kalof and Fitzgerald 2007; Steinberg 2010).

Urban green spaces (parks, treed avenues, gardens, greenbelts, suburban forest patches, greenhearts in city centres) are especially important micro-spaces of animal (especially bird) and human interaction, intersecting the issues of conservation, environmental forestry, recreation, human quality of life and increasing urbanisation (Clergeau et al.1998; Fernandez-Juricic 2000; Park and Lee 2000; Pirnat, 2000; Savard et al. 2000; Marzluff et al. 2001; Chace et al. 2003; Hostetler and Knowles-Yanez 2003; Konijnendijk 2003; Kuhn 2003; Morancho 2003; Chiari and Seeland 2004; Lim and Sodhi 2004; Traut and Hostetler, 2004; White et al. 2005; Campbell 2009c). This issue is unlikely to decline; in fact the issues of recreation and human quality of life, conservation and urban sprawl are increasing (Bird et al.1996; Belant 1997; Jokimaki 1999, 2003; Pirnat 2000; Savard et al. 2000; Cornelis and Hermy 2004; Erickson 2004; Campbell 2010, 2012).

There is little agreement on the impact of such intensified human presence on animals; extinction, adaptation and even proliferation at individual, species and guild levels have been reported (Jokimaki 1999, 2003; Marzluff et al.2001; Konijnendijk 2003 Turner 2003; Vuorisalo et al. 2003; Donnelly and Marzluff 2004; Lim and Sodhi 2004). One argument is that urbanisation may influence clear changes in bird communities; greater species diversity in areas of intermediate urbanisation, and declines with either greater or less urban development (Blair 2004). Studies of avian alert and flight distances in green spaces represent attention to one particular strand in this complex issue: bird reactions to uniform or variable human approaching behavior. This is a narrow focus, as human actions may include movement, seated or standing posture, numbers and food-providing actions (Mavor et al. 2003; Watson 2004). However, this focus provides an important starting point for larger studies, with progressively more animal species, and intra- and inter -species behavior, with the inclusion of human behavior and landcover structure. Then, such studies may inform conservation and green space management, urban development planning and avian ecology, as the building blocks of avian zoogeography (Campbell 2007, 2008a,b, 2009a,b, 2010).

Alert and Flight Distances

Three important parameters for the study of avian reactions to people are the alert distance, the flight distance and the alert period (Burger and Gochfield, 1998; Campbell 2006a, 2010). The alert distance is defined as the distance between the bird and a person at the moment of the bird's first attention to human presence (Fernández-Juricic et al. 2001: 264). The flight distance is the distance between the bird and the approaching person when the bird takes flight. The alert period is the time between the first alert and the bird's flight (Fernández-Juricic et al. 2001). Factors for variation in alert and flight distances include bird size, cover form and configuration, road and other human structure proximity, and human behavior (Campbell 2006).

The Factor of Bird Size

Larger birds are hypothesized to be less tolerant than smaller species of human presence (Cooke 1980; Humphrey et al. 1987; Skagen et al. 1991; Holmes et al. 1993; Gutzwiller et al. 1998). Fernández-Juricic et al. (2001: 264) selected four species: house sparrow (*Passer domesticus,* Linnaeus, 1758), blackbird (*Turdus merula*, L. 1758), wood pigeon (*Columba palumbus*, L., 1758) and magpie (*Pica pica*, L., 1758) as "representative model species" for a study of alert and flight distances. Species size was measured by total length; in this case the house sparrow (15 cm), blackbird (24–25cm), wood pigeon (40–42 cm), and magpie (44–48 cm) (based on Tellería et al. 1999). The larger species (magpie and wood pigeon) were found to be less tolerant (hence, longer alert distances) to human approaches than the smaller species. The blackbird was less tolerant than the smaller house sparrow.

Fernández-Juricic et al. (2001: 267) give several possible reasons for the lower tolerance of human presence from larger species. Firstly, it is possible that "small species have higher energy expenditure than large ones due to greater area/body mass ratios, so small species are expected to be more tolerant of people to diminish energy costs associated with fleeing" (Holmes et al.1993 is cited in this regard). It is also possible that the larger species have a greater ability than smaller species to perceive landscape over a larger radius (a perceptual range) possibly due to sharper vision (see also Martin and Katzir 1994; Lima and Zollner 1996; Kiltie 2000). The authors dispute a suggestion that alert distance variation is related to the costs and benefits (possibly the availability of alternative habitats) of no response, where larger species may have more reason to respond early (Gill et al. 2001).

Campbell (2006a) used 13 species (Table 1) and Campbell (2010) used 10 species Table 2) for similar studies, allowing a greater size range. Campbell (2006a), found very strong correlation between alert distance and bird size. The difference between large and medium-sized birds and the rest, was much greater than the minimal difference between small medium and small species. Campbell (2010) had similar results, with the main difference being between the large corvid and the rest (note: the term 'corvid' refers to relatives of crows, such as rooks, jackdaws, magpies and jays). For the smaller species, there were some inconsistent results. The black-capped chickadee recorded slightly longer alert and flight distances than the slightly larger house sparrow. The eastern phoebe recorded similar alert distances and shorter flight distances to the larger wood thrush.

We might surmise from these studies that human presence may exclude larger bird species in intensely shared spaces. However, this is not the whole story. Other studies I conducted within the field of animal geography found a dominant, individual presence of larger species, especially corvids and doves, when humans provide food (Campbell 2007, 2008). In such cases, human presence (including non-feeding people, possibly indicating anticipatory behavior in birds, as a form of alert period) attracted, rather than repelled birds. Small passerines did not approach people as closely as did the larger species, and medium sized species such as rock doves took refuge near feeding people when attacked by larger crows, a reversal of the findings were human offered food was absent. Human approaches in feeding situations might result in leaving behavior in birds, but this would be only temporary;

anticipatory perching resulted in returns when the person withdrew. Consequently, a more complete picture may emerge habituated birds are compared with less habituated species.

Table 1. Birds in study (Campbell 2006a)

Species	Length (cm)	Wingspan	Weight (g)
Large			
Carrion crow *Corvus corone corone* Linnaeus 1758	45 - 47	93 - 104	370 - 650
Rook *Corvus frugilegus* Linnaeus 1758	44 - 46	81 - 99	280 - 340
Medium			
Magpie *Pica pica*	44 - 46	52 - 60	180 - 270
Jackdaw *Corvus monedula* L.1758	33 - 34	67 - 74	180 - 260
Jay *Garrulus glandarius* L 1758	34 - 35	52 - 58	140 - 190
Small medium			
Blackbird *Turdus merula*	24 - 25	34 – 38.5	80 - 125
Song thrush *Turdus philomelos* Brehm, 1831	23	33 - 36	65 - 100
Starling *Sturnus vulgaris* L. 1758	21.5	37 - 42	60 - 90
Small			
Pied wagtail *Motacilla alba* L. 1758	18	25 - 30	17 - 25
House sparrow *Passer domesticus*	14 - 15	21 – 25.5	24 - 38
Chaffinch *Fringilla coelebs* L. 1758	14.5	24.5 – 28.5	18 - 29
European Robin *Erithacus rubecula* L. 1758	14	20 - 22	14 - 21
Blue tit *Parus caeruleus* L. 1758	11.5	17.5 - 20	9.5 – 12.5

Table 2. Birds in study (Campbell 2010)

Species	Length (cm)	Wingspan	Weight (g)
American Crow *Corvus brachyrhynchos* C.L. Brehm 1822	45	85 - 100	316 - 620
Blue Jay *Cyanocitta cristata* L. 1758	28	34 - 43	70 - 100
American Robin *Turdus migratorius* L. 1766	25	31 - 40	77
Starling *Sturnus vulgaris* L. 1758	22	37 - 42	60- 90
Cardinal *Cardinalis cardinalis* L. 1758	22	25 - 31	42 - 48
Wood Thrush *Hylocichla mustelina* Gmelin 1789	18	31 - 34	40 - 50
Eastern Bluebird *Sialia sialis* L. 1758	17.5	25 - 32	28 - 32
House Sparrow *Passer domesticus* L.1758	16	19 - 25	26 - 32
Eastern Phoebe *Sayornis phoebe* Latham 1790	14	26 - 28	16 - 21
Black-Capped Chickadee *Poecile atricapillus* L. 1766	12	16 - 21	09/14/14

The Factor of Cover

The study by Fernández-Juricic et al. (2001) found that cover proximity and density influenced avian reaction to human presence, with only a small amount of variance. One hypothesis was that the positive correlation of grass cover and alert distances for wood doves and magpies is due to the bird's perception of increased vulnerability to predators in open areas. Denser cover increased blackbird and wood dove tolerance of people, as did shrub height for magpies, possibly due to the perception of lowered predation risk (see also Ydenberg and Dill 1986; Knight and Temple 1995; Martín and López 1995; Kramer and Bonenfant 1997). Tree height also increased alert distances in house sparrows, blackbirds and magpies, possibly due to the increased risk of predation as birds would have to fly a longer distance to reach tree canopies (Fernández-Juricic et al. 2001).

In Campbell (2006a), grass and shrub percentage cover reduced alert distances (this was supported in Campbell 2006b). Larger corvid alert distances were reduced by tree proximity to a greater extent than by shrub proximity. Corvids used larger shrubs with some bare branches for perching after flight in response to an approaching human. Shrub density sharply reduced the alert distances of the smaller species, especially the house sparrow, chaffinch and European robin, which would retreat into especially dense shrubs. The pied wagtail and the blue tit were also more tolerant, but slightly less likely to hide in shrubs, preferring to fly to distant grasses. The house sparrow and chaffinch had significant correlations with understorey, unlike the other small species. The alert distances for the corvids (crows, rooks, jackdaws and magpies) were reduced by understorey proximity.

In Campbell (2010), vegetation height was a significant predictor of alert and flight distances for several species. For crows, shrub height, and tree height (combined with grass and undergrowth), significantly correlated with shorter alert distances, while tree height combined with shrubs was a predictor for longer alert distances. For blue jays, starlings, cardinals, house sparrows, and black-capped chickadees, alert or flight distances were not significantly correlated with vegetation height. For phoebes, only shrub height and tree height with undergrowth were significantly correlated with shorter alert distances. For American robins, wood thrushes and eastern bluebirds shrub height was significantly correlated with shorter alert distances.

The vegetation cover percentage also affected both alert and flight distances, for several species. For crows, increased grass, shrub and tree cover resulted in reduced alert distances, but undergrowth cover had no significant effect. For blue jays, cardinals and phoebes, only shrub cover was linked with shorter alert distances; trees and undergrowth were insignificant as predictors. For starlings and Eastern bluebirds, increased grass and undergrowth cover were linked with shorter alert distances. For house sparrows, tree cover was linked to shorter alert distances, while grass cover was a significant predictor of longer alert distances, and shrub and undergrowth were insignificant in effect. For American robins, wood thrushes and black-capped chickadees no type of vegetation cover was correlated with shorter or longer alert distances.

For the animal geography studies, alert and flight distances were reduced in human proximity, which was invariably in open grassed areas (Campbell 2007, 2008). This subtle alteration of bird behavior indicated that the association of food with people supersede the association of open areas with predation risk. The question emerged as to whether the habituation to feeding in some birds could affect the measurements for alert and flight distances. In these studies, species size was not significantly correlated with human presence and hence vegetation might be unrelated to bird presence; human presence might be the only significant predictor for bird presence.

The Factor of Road and Human Structure Proximity

In Fernández-Juricic et al. (2001) distance from pathways was negatively related to house sparrow alert distance. Consistent with this result, in Campbell (2006a), road and path proximity reduced the alert distances of corvids, starlings, sparrows and chaffinches, but increased the alert distances for the others. In Campbell (2010) road and path proximity were not significant predictors of alert and flight distances for any species. These results encourage the possibility that those species with shorter alert distances near paths and roads are attracted to the possibility of human feeding. This point is not mentioned in Fernández-Juricic et al. (2001), but was a key point in Campbell (2007, 2008). In Campbell (2008) American crows were correlated with roads, paths and pavements; to a lesser extent so were the smaller starlings, rock doves, house sparrows and American robins.

The Factor of Human Behavior

The behavior of the approaching human was a key issue for the alert distances. In Fernández-Juricic et al. (2001: 265), the approaching person walked towards the bird at a steady speed of one step per second; wearing "light blue jeans and grey shirts to avoid differential bird reaction" (see also Gutzwiller and Marcum 1997). In Campbell (2006a) the approach was more complex. On the premise that animals detect human attention towards them, the approaching person's movements were: (1) a straight approach; (2) a sideways approach (walking sideways, with the face at right angles to the bird, using peripheral vision to observe the bird; and (3) a straight approach moving arms by a regular swing up to 45° for each step. The hypothesis was that the sideways approach represented minimal disturbance and hand movements represented maximum movement. Campbell (2010) went further; the first three approaches listed were used and in addition there was a fourth approach where the approaching person swung the arms to a vertical position above the head (90°).

Human directness and aggressive movements (swinging of arms) were decisive factors for longer alert and flight distances for all species. For Campbell (2006a), hand movements particularly affected larger species (corvids); and alert distances doubled or nearly doubled compared with the sideways approach. For smaller species, the increase was between 20 and 30 percent. Alert and flight distances for the straight approach were on average 10 – 35 percent, and 20 to 48 percent shorter than when arm waving was used. For Campbell (2010), there were similar results; arm movements through 90° created the longest alert and flight distances, followed by distances recorded for arm movements through 45°, then those resulting from a straight approach with no arm movement, and finally those resulting from a sideways approach. The largest differences were generally between the straight and sideways approaches with no hand movements and the smallest were generally between the two forms of hand swinging (45° and 90°).

None of these results, however indicate that human presence in parks and shared spaces entirely excludes birds. Campbell (2006b and 2009c) found some species were highly correlated with human presence and modified landscapes compared to more natural landcover. In eastern Canada, these were American crow, starling, American robin and house sparrow. In the UK, such species included the carrion crow, rook, jackdaw, magpie, common starling and house sparrow (personal observation 1993 – 2000, 2003 – 2006). In all cases, highly individual behavior was recorded, including invasion of property and human eating spaces, selective reduction of flight distances to under one meter, observational perching even in open windows and competition for food. In such cases, it was difficult to distinguish alert periods from observational perching.

The Factor of Variable Bird Response

Bird behavior (especially intra- species reactions) in the examined studies was highly variable, as not all the factors for bird presence, alert and flight distances were statistically significant. Fernández-Juricic et al. (2001: 264) acknowledge the possibility of bird adaptation to the approaching technique; hence the approaches were practised in a "20-ha park not included in this study, to avoid conditioning birds in our five study sites." This approach was followed in my studies, as each location selected was new. Bird conditioning to survey methodologies might constitute the genesis of actancy. Avian waiting behavior for anticipated food even in non-feeding situations might also be a factor (personal observations 2007 - 2013).

Despite the lower tolerance of people engaged in hand movements, the studies of human feeding showed that in feeding, hand waving had minimal effect, in some cases encouraging temporary retreats rather than leaving behavior, or observatory perching behavior. As noted above, the alert period might be confused with observational perching, and this might change to inter- or intra-species competition if human presence presented a feeding opportunity. Other human behavior, such as selective feeding of individual birds or species, and aggressive repulsion of some species, did not substantially alter species presence, due to bird persistence and elastic retreat and advance behavior (Campbell 2007, 2008). In fact, one of my studies found some species as positively correlated with human movement (Campbell 2009c).

Avian individualism in response to variable human feeding and non-feeding behavior included waiting behavior of some birds, apparently in hope of success, opportunities created in the feeding spaces and pre- and post-feeding behavior, risk-taking; and the links between this behavior and the variations in human behavior. In the animal geography studies, dominant species present in feeding spaces were Larus gulls, corvids and doves. Smaller passerines, were minimally present; the attraction of large numbers of larger, more social species may also have excluded smaller species. Hence, human behavior may also be relevant not only as an excluding factor for birds, but in attracting aggressive, larger species may also repel or attract individuals of smaller species.

Conclusion

Studies of avian tolerance of human approaches may be useful for environmental planning, especially the establishment of wildlife buffers, which may reduce the disturbance of wildife (Fernández-Juricic et al. 2001). The studies by Campbell (2006a, 2007, 2008, 2010) show that threatening human behviour may also be seen as food providing behavior. Despite the generally lower tolerance of people by larger birds, such species may dominant spaces of feeding or feeding possibilities, which are usually closely shared with people. Consequently, the complementary, more inclusive animal geography model contributes to the thesis of shared actancy, and with alert and flight distance based studies, has applicability to conservation and planning. An informed approach to environmental planning would therefore acknowledge that while certain human behaviors may affect bird presence according to species and size, other human behaviors may also attract birds and in practical situations both repelling and attractive behaviors are likely to be included in shared human-avian spaces. As the animal geography studies showed both inter-species and intra-species differences in feeding in human presence, it is also important to examine inter-species differences in individualist behavior and a broader range of human behaviors within a more inclusive model.

References

Barua, M. 2013. Volatile ecologies: towards a material politics of human-animal relations. *Environment and Planning* A: doi:10.1068/a46138.

Beddard, F.E. 1895. *A Text-Book of Zoogeography.* Cambridge University Press, Cambridge.

Belant, J.L. 1997. Gulls in urban environments: landscape-level management to reduce conflict. *Landscape and Urban Planning,* 38, 245-258.

Bennett, C.F. 1960. Cultural animal geography: an inviting field of research. *Professional Geographer,* 12, 12–14.

Bird, D.M., Garland, D.E. and Negro, J. 1996. *Raptors in human landscapes adaptations to built and cultivated environments*. Academic Press, London.

Blair, R.B. 1996. Land use and avian species diversity along an urban gradient. *Ecological Applications,* 6, 506–519.

Buller, H. 2014. *Animal geographies II Methods.* Progress in Human Geography doi: 10.1177/0309132514527401.

Burger, J. and Godchild, M. 1998. Effects of eco-tourists on bird behaviour at Loxahatchee National Wildlife Refuge, Florida. *Environmental Conservation* 25 13–21.

Cansdale, G.S. 1952. *Animals and Man Hutchinson,* London.

Campbell, M., 1998. *Interactions between biogeography and rural livelihoods in the coastal savanna of Ghana.* Unpublished Ph.D. Thesis, University of London.

Campbell, M. 2006a. Urban parks as shared spaces? The utility of alert distances as indicators of avian tolerance of humans in Stirling, Scotland. *Area* 38(3): 301- 311.

Campbell, M. 2006b. *Impact of Landscape Features on Corvids in Stirling and Lewes.* Unpublished Certificate Thesis, University of Lancaster, Lancaster.

Campbell, M. 2007. An animal geography of avian ecology in Glasgow. *Applied Geography* 27(2): 78-88.

Campbell, M. 2008. An animal geography of avian feeding habits in Peterborough, *Ontario Area.* 40(4): 472–480.

Campbell, M. 2009a. A new zoogeography of domestication and agricultural planning in Southern Ghana. *Area* 41(2): 139–148.

Campbell, M. 2009b. Repositioning zoogeography within the nature–culture borderlands: An animal geography of reptiles in southern Ghana. *Applied Geography* 29: 260–268.

Campbell, M. 2009c. The impact of habitat characteristics on bird presence and the implications for wildlife management in the environs of Ottawa, Canada. *Urban Forestry & Urban Greening* 8(2): 87–95.

Campbell, M. 2010. Passerine reactions to human behaviour and vegetation structure in Peterborough, Canada. *Urban Forestry & Urban Greening* 10 47–51.

Chace, J.F., Walsh, J.J., Cruz, A., Prather, J.W., Swanson, H.M. 2003. Spatial and temporal activity patterns of the brood parasitic brown headed cowbird at an urban/ wildland interface. *Landscape and Urban Planning* 64: 179–190.

Chiari, C.G., Seeland, K., 2004. Are urban green spaces optimally distributed to act as places for social integration? Results of a geographical information system GIS ap- proach for urban forestry research. *Forest Policy and Economics* 6, 3–13.

Clark, C. W. 1973. Profit maximization and the extinction of animal species. Journal of Political Economy 81: 950-961.

Clergeau P, Savard J P L, Mennechez G, Sauvage A and Falardeau G 1998 Bird abundance and diversity along an urban–rural gradient: a comparative study between two cities on different continents. *Condor* 100 413–25.

Cooke, A.S. (1980) Observations on how close certain passerine species will tolerate an approaching human in rural and suburban areas. *Biological Conservation* 18: 85–88.

Cornelis, J., Hermy, M., 2004. Biodiversity relationships in urban and suburban parks in Flanders. *Landscape and Urban* Planning 69, 385–401.

De Angelis, D.L and Waterhouse, J.C. 1981. Equilibrium and nonequilibrium concepts in ecological models. *Ecological Monographs* 57: 1-21.

De Leo, G. A., and Levin, S., 1997. *The multifaceted aspects of ecosystem integrity. Conservation Ecology* [online]1(1), 3. Available from the Internet. URL: http://www.consecol.org/vol1/iss1/art3

Donnelly, R. and Marzluff, J.M. 2004. Importance of reserve size and landscape context to urban bird conservation. *Conservation Biology* 18: 133–145.

Drude 1884. Die Florenreiche der Erde in 'Petermann's Mitteilungen'.

Drude, C.G.O. 1884. Die Florenreiche der Erde: Darstellung der gegenwärtigen Verbreitungsverhältnisse der Pflanzen. *Petermanns Mitteilungen aus Justus Perthes' Geographischer Anstalt,* Erganzbd 15(74).

Emel, J. and Urbanik, J. 2010. Animal geographies: exploring the spaces and places of human-animal encounters. In DeMello, M. (ed.) *Teaching the Animal: Human-Animal Studies across the Disciplines,* 200-217. Lantern Books, New York.

Emel, J., Wilbert, C. and Wolch, J. 2002. Society and animals. *Journal of Human Animal Studies,* 10(4), 1.

Fernández-Juricic, E., Dolores, M., Lucas, E., 2001. Alert distance as an alternative measure of bird tolerance to human disturbance implications for park design. *Environmental Conservation* 28: 263–269.

Ganderton, P. and Coker, P. 2005. *Environmental Biogeography.* Pearson/Prentice Hall, Harlow.

Gill, J.A., Norris, K. and Sutherland, W.J. 2001 Why behavioural responses may not reflect the population consequences of human disturbance. *Biological Conservation* 97: 265– 8.

Gutzwiller, K.J. and Marcum, H.A. 1997. Bird reactions to observer clothing color: implications for distance-sampling techniques. *Journal of Wildlife Management* 61: 935–947.

Gutzwiller, K.J., Marcum, H.A., Harvey, H.B., Roth, J. and Anderson, S.H. 1998. Bird tolerance to human intrusion in Wyoming Montane Forests. *Condor* 100: 519–527.

Heilprin, A. 1887. The Geographical and Geological Distribution of Animals. *International Sci. Series.* 12mo. D.Appleton & Co., New York.

Hodgetts, T. and Lorimer, J. 2014. Methodologies for animals' geographies: cultures, communication and genomics. Cultural Geographies. doi: 10.1177/1474474014525114

Holling, C.S. 1973. Resilience and Stability of Ecological Systems. *Annual Review of Ecological System* 14: 1-24.

Holling, C.S. 1993. Investing in Research for Sustainability" *Ecological Applications.* 3: 552-555.

Holmes, T.L., Knight, R.L., Stegall, L. and Craig, G.R. 1993. Responses of wintering grassland raptors to human disturbance. *Wildlife Society Bulletin* 21: 461–468.

Hostetler, M. and Knowles-Yanez, K. 2003. Land use, scale and bird distributions in the Phoenix metropolitan area. *Landscape and Urban Planning* 62: 55–68.

Hovorka, A. 2008. Transspecies urban theory: chickens in an African city. *Cultural Geographies* 15: 95–117

Humphrey, P.S., Livezey, B.C. and Siegel-Causey, D. 1987. Tameness of birds of the Falkland Islands: an index of preliminary results. *Bird Behaviour* 7: 67–72.

Ingersoll, E. 1920. *The Encyclopedia Americana* (1920)/Zoogeography.

Ingersoll, E. 1920. Zoogeography. In the Encyclopedia Americana. Grolier, New York.

Johnston, R. J., Gregory, D., Pratt, G. and Watts, M. 2000. *Human geography* (4th ed). Blackwell Publishing, Malden.

Jokimaki, J. 1999. Occurrence of breeding bird species in urban parks: effects of park structure and broad-scale variables. *Urban Ecosystems* 3: 21–34.

Kalof, L. and Fitzgerald, A.J. (eds.) 2007. *The Animals Reader: The Essential Classic and Contemporary Writings.* Berg, New York.

Karr, J.R. and Dudley, D.R. 1981. Ecological perspective on water quality goals. *Environmental Management* 5: 55-68.

Kiltie, R.A. 2000. Scaling of visual acuity with body size in mammals and birds. *Functional Ecology* 14: 226–234.

Knight, R.L. and Temple, S.A. 1995. Wildlife and recreationists: coexistence through management. In Knight R.L. and Gutzwiller, K.J. (eds.) *Wildlife and Recreationists: Coexistence Through Management and Research,* 327–333. Island Press, Washington, DC.

Konijnendijk, C.C. 2003.A decade of urban forestry in Europe. *Forestry Policy and Economics* 5:173–186.

Kramer, D.L. and Bonenfant, M. 1997. Direction of predator approach and the decision to flee to a refuge. *Animal Behaviour* 54: 289–295.

Kuhn, M. 2003. Greenbelt and greenheart: separating and integrating landscapes in European city regions. *Landscape and Urban Planning* 64: 19–27.

Lim, H.C. and Sodhi, N.S. 2004. Responses of avian guilds to urbanisation in a tropical city. *Landscape and Urban Planning* 66: 199–215.

Lorimer, J. 2007. Nonhuman charisma. *Environment and Planning D: Society and Space* 25, 911-932.

Ludwig, D., Hilborn, R. and Walters, C. 1993. Uncertainty, resource exploitation, and conservation: lessons from history. *Science* 260: 17, 36.

Marzluff, J.M., Bowman, R. and Donnelly, R. (eds.) 2001. *Avian Ecology and Conservation in an Urbanising World Kluwer.* Academic, Boston.

Martin, G.R. and Katzir, G. 1994. Visual fields and eye movements in herons (Ardeidae). *Brain, Behaviour, and Evolution* 44: 74–85.

Martín, J. and López, P. 1995. Influence of habitat structure on the escape tactics of the lizard Psammodromus algirus. *Canadian Journal of Zoology* 73: 129–132.

Mavor, R.A., Parsons, M., Heubeck, M., Pickerell, G. and Schmidtt, S. 2003. *Seabird numbers and breeding successin Britain and Ireland 2002 Joint Nature Conservation Committee,* Peterborough.

McDonald, G. 2003. *Biogeography: Introduction to Space,* Time and Life. John Wiley and Sons, New York.

Morancho, A.B., 2003. A hedonic valuation of urban green areas. *Landscape and Urban Planning* 66: 35–41.

Murdoch, J. 1998. *The spaces of actor network theory Geoforum* 29: 357–74

Nicolis, G. 1994. Dynamical systems, biological complexity and global change. In Solbrig, O.T., Van Emden, H.M. and Oordt, P.G.W.J., *Biodiversity and Global Change,* 21-31, CAB International, Wallingford.

Oilman, D.C., Peck, H.T. And Colby, F.M. 1905. *The New International Encyclopædia*/Distribution of Animals. Dodd, Mead and Company, New York.

Park, C. R. and Lee, W. S. 2000. Relationship between species composition and area in breeding birds of urban woods in Seoul Korea. *Landscape and Urban Planning* 51: 29–36.

Philo, C. and Wilbert, C. (eds.) 2000. *Animal spaces beastly spaces: new geographies of human–animal relations.* Routledge, London.

Pickett, S.T.A., Parker, V.T. and Fiedler, P.L. (1992) "The New Paradigm in Ecology: Implications for Conservation Biology above the Species Level" in Fiedler, P.L. and Jain, K., (1992) *Conservation Biology: The Theory and Practice of Nature Conservation, Preservation and Management* pp.66-87 Chapman and Hall, NewYork.

Pirnat, J., 2000. Conservation and management of forest patches and corridors in suburban landscapes. *Landscape and Urban Planning* 52, 135–143.

Rowe, J.S., 1997. From reductionism to holism in ecology and deep ecology. *The Ecologist* 27 (4), 147–151.

Savard, J.P.L., Clergeau, P., Mennechez, G., 2000. Biodiversity concepts and urban ecosystems. *Landscape and Urban Planning* 48, 131-142.

Sclater, P.L. 1858. *Journal of the Proceedings of the Linnean Society: Zoology* 2 (1858): 130-145.

Scoones, I., 1997. The dynamics of soil fertility change: historical perspectives on environmental transformation from Zimbabwe. *The Geographical Journal* 63 (2), 161–169.

Skagen, S.K., Knight, R.L. and Orians, G. 1991. Human disturbance of an avian scavenging guild. *Ecological Applications* 1: 215–225.

Steinberg, M.K. 2010. Avifaunal research and Geographical Perspectives. *Geographical Review* 100(3): iii–iv.

Stott, P.A. 1994. Savanna landscapes and global environmental change. In Roberts, N. 1994. *The Changing Global Environment,* 287-303. Basil Blackwell Limited, Cambridge.

Sullivan, S., 1996. Towards a nonequilibrium ecology: perspectives from an arid land. *Journal of Biogeography* 22: 1–5.

Tellería, J.L., Asensio, B. and Díaz, M. 1999. *Aves Ibéricas. II. Paseriformes.* J.M. Reyero, Madrid.

Traut, A.H. and Hostetler, M.E., 2004. Urban lakes and waterbirds effects of shoreline development on avian distribution. *Landscape and Urban Planning,* 69: 69–85.

Urbanik, J. 2012. *Placing Animals: An Introduction to The Geography of Human-Animal Relations.* Rowman & Littlefield, Lanham.

Vuorisalo, T., Andersson, H., Hugg, T., Lahtinen, R., Laaksonen, H. and Lehikoinen, E. 2003. Urban development from an avian perspective: causes of hooded crow Corvus corone cornix urbanisation in two Finnish cities. *Landscape and Urban Planning* 62: 69–87.

Wallace, A.R. 1876. *The Geographical Distribution of Animals: With a Study of the Relations of Living and Extinct Faunas as Elucidating the Past Changes of the Earth's Surface,* 2 Vols. Macmillian, London, Harper and Brothers, New York.

Watson, J. 2004. *Gulls strike terror in the streets* (http://www.scotlandonsunday.com) Accessed 16 July 2013.

Weins, J. 1984. On understanding a non-equilibrium world: myth and reality in community patterns and processes. In Strong, D.R. (ed.) *Ecological Communities: Conceptual Issues and the Evidence,* 439-457. Princeton University Press, Princeton.

Wolch, J. and Emel, J. (eds.) 1998. *Animal geographies: Place, Politics and Identity in the Nature-Culture Borderlands.* Verso, London.

Yarwood, R. and Evans, N. 2000. Taking stock of farm animals and rurality. In Philo, Cand Wilbert, C. (eds.), Animal spaces, Beastly Spaces. *New Geographies of Human-animal Relations,* 98–114). Routledge, London.

Ydenberg, R.C. and Dill, L.M. 1986. The economics of fleeing from predators. *Advances in the Study of* Behaviour 16: 229–249.

Young, M.D. and Solbrig, O.T., 1993. Providing an environmentally sustainable, economically profitable and socially equitable future for the world's savannas. In Young, M.D. and Solbrig O.T. (eds.), *The World's Savannas: Economic Driving Forces, Ecological Constraints and Policy Options for Sustainable Land Use,* 321 – 344. Pantheon Publishing Group, Paris.

Zimmerer, K.S. 1994. Human geography and the 'New Ecology': the prospect of integration. *Annals of the Association of American Geographers* 84(1): 108-125.

In: Advances in Zoology Research. Volume 6
Editor: Owen P. Jenkins
ISBN: 978-1-61122-061-2

Chapter 5

A Comparative Analysis of Human Perceptions of Cougars and other Large Carnivores in Canada and El Salvador

Michael O'Neal Campbell*

Lakehead University, Thunder Bay, Ontario, Canada

Abstract

The relationship between people and the cougar *Puma concolor* and other larger predators in Canada (brown bears *Ursus arctos*, black bears *Ursus americanus*) and Latin America (jaguars *Panthera onca*) is examined in a comparative mode for cases in Kamloops and Vancouver Island, Canada and El Salvador. Literature derived hypotheses mooting greater sympathy for carnivores from women and young people, greater fear from women and older people, and greater fear of the larger brown bears and jaguars over cougars are tested. The results show that the hypotheses are both supported and refuted. The size of the predator is as important as common ideas about its behaviour, and troublesome invasive behaviour (mostly from bears) is at least as important as predatory behaviour (largely from cougars and jaguars). However, consistent with hypotheses, women and older people were in many instances more afraid of cougars and other carnivores, and older people were generally less tolerant of carnivores. Differences included opinions on positive impacts on quality of life and useful functions of cougars (favoured by men in El Salvador, both men and women in Vancouver Island, and women in Kamloops), and danger to adults, children and other animals (favoured by women more than men in El Salvador for all groups, women more than men in Vancouver Island for children and animals, not adults, and women more than men in Kamloops for adults and children, not animals). For age, older people in El Salvador were generally less tolerant than younger people of large carnivores, similar to Vancouver Island, but not Kamloops, where there were no age-related opinions of the danger from carnivores towards people. The cougar's elusive behaviour limited comparisons with bears and jaguars. Awareness of such issues may assist wildlife conservation.

* Email: ecol55@hotmail.com.

INTRODUCTION

Large, wild carnivores hold a special and unique position in the conservation debate, because they contribute to strongly negative attitudes among people, especially in rural areas, due to fear of attacks on humans and/or livestock, resultant restrictions on human life spaces, and the monetary, life and lifestyle losses incurred as a result (Saenz and Carrillo 2002; Conforti et al. 2003; Kleiven et al. 2004; Hoogesteijnal & Hoogesteijnal 2008). The study of large carnivores is also of practical relevance, because wide-ranging nocturnal carnivores are frequently used as umbrella species in conservation planning (Soule & Terborgh 1999; Maehr et al., 2001, 2002; Carroll et al. 2003). As such their relations with people influence habitat and landuse planning, and conservation policy (Dickson & Beier 2007). Debateable issues concern the actual inter-species based factors for human attitudes towards particular species, and the relevance of these attitudes for conservation possibilities (Campbell and Lancaster 2010). In all continents, different species of large carnivores share landscapes and spaces of conflict with people, with examples being: brown bears *Ursus arctos* and wolves *Canis lupus* in Eurasia (Schwartz et al. 2003); tigers *Panthera tigris* and leopards *Panthera pardus* in India and South east Asia (Johnsingh 1983; Karanth & Sunquist 1995; Johnsingha & Negib 2003); lions *Panthera leo,* leopards and cheetahs *Acinonyx jubatus* in sub-Saharan Africa (Maclennan et al. 2009); cougars *Puma concolor*, black bears *Ursus americanus* and brown bears *Ursus arctos* in North America (Kellert et al. 1996; Decker et al. 2001); and cougars and jaguars in Latin America (Taber 1997; Scognamillo et al. 2003; Campbell 2011).

In such scenarios of conservation, conflict and strategic possibilities, important issues concerning inter-species variables are the aggressiveness, perceived or real, population density, landcover preferences, size, and reputation of the resident carnivores, these being important factors for public attitudes and hence conservation efforts (Treves & Karanth 2003; Hayward & Somers 2009; Inskip & Zimmermann 2009; Campbell & Lancaster 2010). For example, in North America several studies have found that due to interspecies differences in size and behaviour, people generally have more positive opinions about black bear presence than that of the cougar, because of the perception that the latter animal is more predatory (Davis et al. 2001; Beier, 1991). Brown bears are more feared than black bears, because of their larger size and reputed aggressiveness (Kellert, 1994; Pelton et al. 1976). Furthermore, people are more tolerant of black bears than of cougars and brown bears, and of cougars than brown bears (Pelton et al. 1976; Hayward & Somers, 2009; Deurbrouck, 2007; Thornton & Quinn, 2009; Kellert, 1994) Concerning human variables, in the comparative analysis of human attitudes to two or more species of predators, the respondents' gender and age are particularly important (Campbell & Lancaster 2010). Several hypotheses regarding gender and age related opinions towards large carnivores have been posed in the literature. As argued by Weber & Rabinowitz (1996), Kleiven et al. (2004) and Campbell & Lancaster (2010), women may be more tolerant of large carnivores than men, but express more fear and more concern for child safety.

These hypotheses have been examined in the North American context (Morzillo et al. 2007; Bowman et al. 2001, 2004; Decker et al. 2001; Kellert et al. 1996). Kellert (1994) and Williamson (2002) hint that demographic characteristics such as age and gender may contribute to different attitudes toward large carnivores, with women and older people being more concerned because they feel more vulnerable. Women are more afraid of bears than men are of these animals (Morzillo et al. 2010). Women and younger people are more likely than men and their elders, to support carnivore conservation by choosing toleration or trapping and removal rather than shooting the animals (Fulton et al. 1995; Kellert, 1976; 1989; Morzillo et al. 2007; Teel et al. 2002; Hayward & Somers, 2009; Deurbrouck, 2007; Thornton & Quinn, 2009).

The Case of the Cougar *Puma concolor* and the Larger Sympatric Carnivores

Few carnivores have as wide a range as the cougar, also known as the puma, mountain lion and panther, this species being found through most of North and South America (Campbell & Lancaster 2010). Throughout its range, the cougar shares its habitats with people, and also with larger carnivores (brown bears *Ursus arctos* and black bears *Ursus americ*anus in North America, and jaguars *Panthera onca* in Mexico southwards (Taber 1997; Nunez et al. 2000, 2002; Decker et al. 2001; Crawshaw & Quigley, 2002; Oliveira, 2002; Campbell & Lancaster 2010, Campbell & Torres Alvarado 2011). The cougar is generally perceived as an elusive, secretive carnivore, with few records of attacks on people, moderate danger to livestock, but fear of attacks on people, mostly children (Beier, 1991, 1993; Turner et al. 1990; Herbert, 1989). Cougars, similar to the larger carnivores, are frequently situated in scenarios of conflict with people, due to human population expansion, landscape change and habitat conversion, increased livestock presence and perceived danger (Campbell & Torres Alvarado 2011). These conflicts are especially important for carnivore survival, because human comparisons of different species strongly affect the conservation possibilities of different species (Treves & Karanth 2003; Hayward & Somers 2009; Inskip & Zimmermann 2009).

There are both positive and negative attitudes towards cougars, the positive views based mainly on the cougar's status as a wildlife symbol and its relation to the domestic cat (Deurbrouck, 2007; Thornton & Quinn, 2009), and the negativity based on their predation threat to pets and children (McKee, 2003). Few studies have compared the public attitudes towards cougars in widely distributed parts of its range, and included a comparison with other predators in these varied regions, although some studies have compared public attitudes and ecological relations of cougars and bears in North America (Kellert et al. 1996; Decker et al. 2001; Teel et al. 2002) and similar topics on cougars and jaguars in Central and South America (Jorgenson & Redford 1993; Taber et al. 1997; Farrell 1999; Polisar 2000, 2003; Conforti et al. 2003; Scognamillo et al. 2003; Monroy-Vilchis 2009).

Key questions and hypotheses that would emerge from such a comparative study (necessarily based on several literature sources, as the vastness of the cougar's range would make a field study difficult) would concern the importance of the animals' physical size (in this example, the cougar is generally smaller than the bears and the jaguar), and behaviour, as the cougar is generally seen as more elusive (even though the jaguar and brown bear may generally be reported as rarer), less aggressive than the other animals, less of a threat to livestock and humans, except to children, and less invasive than bears. Also important would be the human variables, generally gender and age, as these are all pervasive, and in some cases education and employment, even though these latter variables may be less important in the rural contexts where agriculture is the main livelihood activity. This paper analyses the results of three articles that attempt a comparison of cougars, bears and jaguars (Campbell & Lancaster 2010, Campbell & Torres Alvarado 2011 and Campbell 2012) based on Vancouver Island and Kamloops in Canada and El Salvador in Central America, and evaluates their findings within the broader context of the literature on the specific species, the sociology of human gender and age and general human relations with large carnivores.

THE COUGAR AND THE JAGUAR

A major factor for the different and similar attitudes to jaguars and cougars is their size, and consequently the size and type of prey taken by these cats (Campbell and Torres Alvarado 2011). Jaguars are described as the third largest cat species (after tigers and lions), with some male jaguars weighing as much as female tigers or lions (up to 160 kg) especially in the Amazon forests, but averaging 56–96 kg (Burnien & Wilson 2001). Cougars are described as the fourth largest cats, marginally larger than leopards, but varying greatly in size (for example cougars in the colder regions of Canada are described as larger than some jaguars (75 kg) (Polisar 2000, 2003). Their size range is about 29 - 100 kg (males typically weigh 53 to 100 kilograms) (Nowell & Jackson 2006). Strong positive correlations have been described between mean prey mass and cougar (Iriarte et al. 1990) and jaguar size (Oliveira 1994). According an analysis by Oliveira (1994) the jaguar:cougar body ratio was more equal in Central (1:0.9) than South America (1:0.6), similar to the mean prey weight ratio in Central (1:0.9) and South America (1:0.6), hinting that cat size affects the size of prey taken. Studies have found that jaguars kill larger mammals than cougars, in some cases the prey of the jaguar weighing twice as much as that of the cougar (Oliveira. 2002; Palmeira et al. 2008).

These characteristics provide evidence for the relations between these cats and human life spaces. In terms of domestic animals, cattle have been described as important targets, but jaguars may focus on adult animals while cougars take calves, and smaller animals such as sheep and goats (Yàñez et al. 1986; Boulhosa and Valdes, 1997; Oliveira. 2002). Neither the jaguar nor the cougar commonly kills people (unlike tigers, lions and leopards in other continents), despite fears to the contrary in North America for the cougar and mostly South America for the jaguar (Conforti et al. 2003. Campbell & Lancaster 2010, Campbell & Torres Alvarado 2011). In areas of joint distribution of cougars and jaguars, the latter cat is more feared, more blamed for livestock and human attacks and more known, even when as noted by Conforti et al. (2003) these judgements are not justified by evidence (Perry 1970; Guggisberg 1975; Almeida 1990; Beier 1991; Foerster 1996). The cougar is less known than the jaguar, due to its elusive nature (Conforti et al. 2003; Campbell 2010).

The results of my study (Campbell & Torres Alvarado 2011) of jaguars and cougars in El Salvador are particularly interesting, as these animals are largely extinct in that country (but appear to cross borders), and hence reintroductions are an important topic. The jaguar and cougar remain in the public consciousness due to livestock kills and threats to people (Conforti and Azevedo 2003). The relatively small size of El Salvador facilitates animal movements and prevents final judgements on extinction. El Salvador (population 6,052,064, area 21,041 square kilometres, population density 288 per square kilometre; Central Intelligence Agency 2010, Figure 1) is the most densely populated country in Central and

South America (Hecht et al. 2006). In this study four hypotheses were posed, based on the literature on human attitudes to these animals (Weber and Rabinowitz 1996; Kleiven et al. 2004; Campbell and Lancaster 2010). These were: (1) women are more tolerant of large carnivores than men; (2) women are more afraid of large carnivores; (3) both men and women are more afraid of jaguars than cougars; and (4) older people are less tolerant of large carnivores than younger people. For a comparison of the cougar and the jaguar, the most relevant was hypothesis 3.

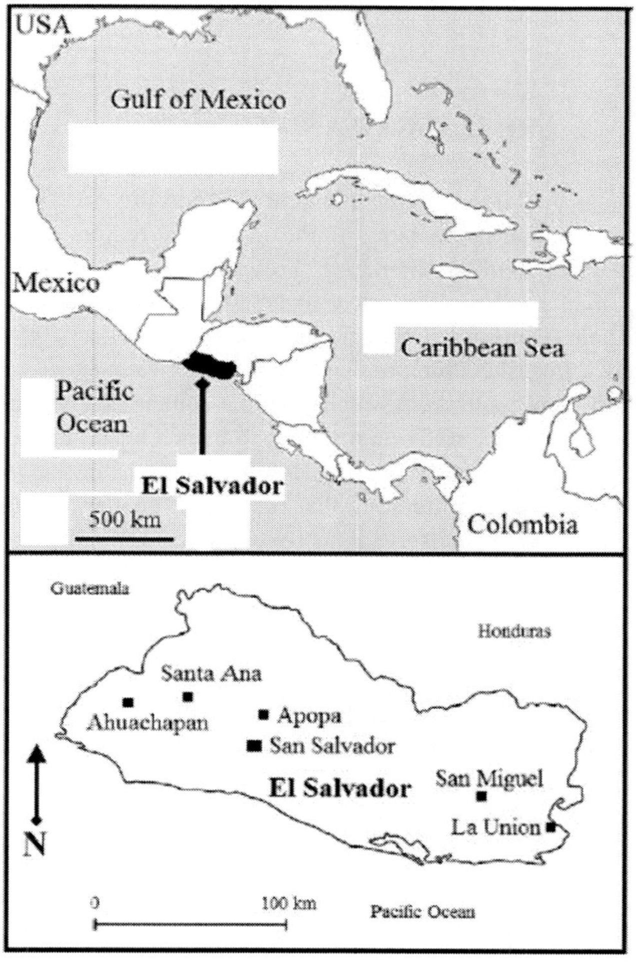

Figure 1. Location of El Salvador.

The results showed that there were no significant differences in attitudes towards the two species in terms of toleration, trapping and removal, shooting, protection, reintroductions, impacts on human life, or danger to cattle, human adults or children. As explained by some respondents, any difference in strength between the jaguar and cougar would make little difference to a man, woman or child in an encounter. However, more people thought cougars were more dangerous than jaguars (especially women, although this did not affect the other assessments). This runs against the common perception of the jaguar as a powerful cattle killer, and the cougar as the more elusive, timid animal, in some cases not even dangerous to people (Conforti and Azevedo 2003; Palmeira et al. 2008). It nevertheless supports the findings of the study by Conforti and Azevedo (2003) on the comparative local fears and perceptions of jaguars and cougars, which found that although fewer people feared the cougar than the jaguar, feelings towards the two species were similar: people with positive perceptions towards jaguars felt the same towards cougars, and people with a negative opinion of jaguars tended to have the same opinion of cougars, regardless of actual behavioural differences between the two cats.

THE COUGAR AND THE BROWN BEAR

The brown bear is the largest mammalian carnivore in the Americas, weighing from 100 to 322 kg, with most of the larger bears found in coastal regions of British Columbia and Alaska (McLellan 1994). In the literature, brown bears are more feared, with a more 'nefarious image,' than cougars (Kellert et al.1996)., and are also more feared than black bears, because of their superior size and fierce reputation (Kellert 1994; Pelton et al. 1976), despite the generally positive opinion that North Americans have of bears (Hayward & Somers 2009; Kellert 1994), and the perception of the cougar as a predation threat to pets and children (McKee, 2003). In my study on attitudes to brown bears, black bears and cougars in Kamloops, British Columbia (Campbell, forthcoming for 2012 in the *Journal of Society and Animals,* Figure 2), based on the literature that found that gender and age related differences in support for bear and cougar conservation (Teel et al. 2002) and fear of bears and cougars (Kellert, 1989, 1994; Thornton & Quinn, 2009; Morzillo et al. 2010) the following hypotheses were posed: (1) women and older people are more afraid of bears and cougars than men and younger people, respectively; (2) women and younger people are more likely than men and their elders, respectively, to support bear and cougar conservation by choosing toleration or trapping and removal rather than shooting the animals; and (3) more people will be tolerant of black bears than of cougars and brown bears, and of cougars than brown bears. The evidence for the third hypothesis, showed that cougars were seen as the most dangerous to children and pet animals, followed by brown bears. Brown bears were seen as the most dangerous to adults, followed by cougars.

THE COUGAR AND THE BLACK BEAR

The black bear is generally larger than the cougar (41 - 250 kg) (Brown 1993). In areas of joint distribution, cougars are perceived as more dangerous than black bears, the main indicators being that more people (especially men) approve of cougar hunting than black bear hunting (Teel et al. 2002). In my paper on black bears, cougars and brown bears in Kamloops (Campbell 2012), black bears were seen as the least dangerous of the three animal species. In my study comparing cougars and black bears in Vancouver Island, Canada (Campbell & Lancaster 2010, Figure 3), based on the findings of the literature on public perceptions of black bears and cougars in North America (Morzillo et al., 2007; Bowman et al.,2001, 2004;

Decker et al., 2001; Kellert et al., 1996), the following hypotheses were investigated: (a) people generally have positive more opinions about black bear than cougar presence, because of the perception that the latter animal is more predatory (Davis et al. 2001; Beier 1991); (b) demographic characteristics such as age and gender may contribute to different attitudes toward large carnivores, with women and older people being more concerned because they feel more vulnerable (Williamson 2002; Kellert 1994).

In terms of the first hypothesis, few responses indicated any species-specific differences in people's perceptions, suggesting that people view the two predators similarly. Cougars were perceived as more dangerous overall than bears, especially toward children and companion/farm animals, but bears were perceived as engaging in more troublesome behaviours than cougars. Attitudes about how to treat the two predators were essentially identical, but there were differences on the reasons. Reasons for shooting cougars comprised the perception of their predatory nature (compared with the more omnivorous bears) and their greater likelihood to attack pets (dogs, cats, and birds), livestock, and children. Reasons for shooting bears mostly concerned. their invasion of private property or social spaces. Similar to these findings, Morzillo et al. (2007) point out that, at a general level "people tend to respond quite similarly to different large carnivore species regardless of ecological and behavioural differences" (p. 418). This point is supported by several studies (Kleiven et al., 2004; Kellert et al., 1996; Kellert, 1985).

THE IMPORTANCE OF GENDER

According to literature sources, North American women are more afraid than men of cougars (Kellert, 1989, 1994; Thornton & Quinn, 2009) yet women are more supportive than men of cougar conservation (Kellert, 1989; Teel et al. 2002; Thornton & Quinn, 2009). Based on the results in my paper on El Salvador (Campbell & Torres Alvarado 2011), there were important findings based on gender. More women than men were afraid of the big cats. More men than women supported toleration of jaguars and cougars, with no gender differences on trapping and shooting, similar proportions of men and women thought the cats needed more protection, or could be reintroduced into protected areas, general rural areas or zoos. Slightly more women than men saw a danger to pets, livestock and people, but this was not statistically relevant. More men than women thought jaguars were more dangerous, while due to undecided answers more men than women also thought that cougars were more dangerous than jaguars, but by a smaller margin. More women than men thought the presence of the carnivores made human life worse, and more men than women thought they made life better. Relating these results to the hypotheses, the first hypothesis of the study was that women are more tolerant of large carnivores than men. This hypothesis is supported by eco-feminist, 'women and environment' and the political ecology literature that women, through cultural conditioning and/or natural cognition have a greater sympathy for the environment/wildlife (Campbell 1998, 2005a, 2005b). This hypothesis was generally rejected, as men were more tolerant of large carnivores, and gender differences in the shooting of the animals were insignificant. Men were also more likely to agree that large carnivores improved human life, while more women thought otherwise. For the second hypothesis, that women are more afraid of wildlife and are more concerned about the danger to children, the results supported the first part but not the second (in contrast to papers on Canada described below).

Figure 2. Kamloops, in Canada.

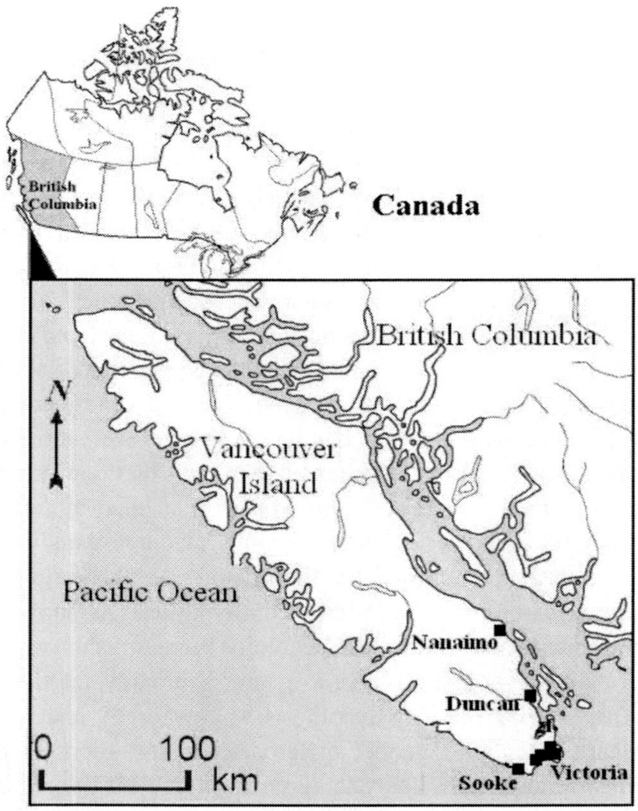

Figure 3. Vancouver Island, in Canada.

Based on the hypotheses of my two Canadian papers, there were also important findings based on gender. In the study on Vancouver Island (Campbell & Lancaster 2010), while men and women differed little on whether to shoot, trap or remove cougars and black bears, more women than men thought that carnivores were dangerous to pet animals and livestock and children. Men and women did not differ on the danger to adults, although more people saw cougars as dangerous than bears. While people thought bears were more likely to engage in troublesome behaviour than cougars, more men than women thought that bears were troublesome (there was no gender distinction for cougars). Most people did not think that carnivores diminished the quality of life, and there were no gender distinctions. In my second study for Kamloops (Campbell 2012), significantly more women than men viewed the carnivores as dangerous to adults and children, partly because many women believed that women were closer to children. In this study, more women than men thought the bears were troublesome, but there was little difference between women and men in their opinion of cougars. There were no differences for gender regarding shooting, trapping or removal of the carnivores. More women than men thought cougars had useful functions, and enhanced human quality of life, but their opinions about bears were similar.

The Importance of Age

According to the literature, younger people are more supportive of cougar conservation and less afraid of these animals than their elders (Fulton et al. 1995; Kellert, 1976; 1989; Teel et al. 2002). Concerning the hypothesis about age in the study of El Salvador (Campbell & Torres Alvarado 2011), the fourth hypothesis was that older people would be less tolerant of wild animals than younger people. However, age was irrelevant to the fear of the animals, toleration, reintroductions to unrestricted rural areas and zoos, but there was greater support among older people for shooting and less support for reintroductions to special areas and a stronger perception of danger to cattle. These findings are partly consistent and inconsistent with other literature on the topic that young people are more tolerant of large carnivores (Andersone & Ozolins 2004) and of environmental issues in general (Batterbury et al. 1997; Campbell 2005b). For the study of Vancouver Island (Campbell & Lancaster 2010), as noted above, the second hypothesis was that human age could contribute to different attitudes toward large carnivores, with women and older people being more afraid because they feel more vulnerable (Williamson 2002; Kellert 1994). The results showed that more younger people and fewer older people supported tolerance of large carnivores, and fewer younger people and more older people supported the trapping and removal of the animals. Although more older people saw bears as a dangerous to pets and livestock, and the reverse was true for danger to adults, there were no significant differences between young and old for cougars, or for the danger of any animal to children. Neither older or younger people believed that cougars were troublesome, while more younger people thought that bears were more troublesome than cougars, and as a consequence were less likely to reduce the quality of life.

For the study of Kamloops (Campbell 2012) the first two hypotheses were relevant to age (older people will be more afraid of bears and cougars, and younger people are more likely to support bear and cougar conservation by choosing toleration or trapping and removal rather than shooting the animals). The results showed that there were no age-related opinions of the danger from carnivores towards people, but more older people saw brown bears as dangerous towards pets. Also, there were no age-related opinions on the troublesome behaviour of cougars, but older people were more likely than younger people to see bears as troublesome. Inconsistent with hypothesis 2, and although most people selected trapping and removal over shooting, more younger people believed cougars and bears should be shot, and more older people believed in trapping and removal. Age-related opinions on the usefulness, and contribution to quality of life, of cougars and bears were insignificant.

Comparisons and Conclusion

The results of these three papers, situated within the broader literature review illustrate some important points. Firstly, the cougar was seen as a dangerous animal, only slightly different from the larger carnivores. While broadly similar to the jaguar and brown bear in dangerousness, it was definitely more dangerous, at least in some situations than the black bears. Its potential as a predator appeared more important than its elusive behaviour, the latter characteristic only matched by the troublesome behaviour of bears. Secondly, gender was a very important variable affecting viewpoints, but these were inconsistent across the study areas. In all three cases, more women than men thought the carnivores were dangerous, especially to people but also to other animals, but in terms of the lesser problem of troublesome behaviour and positive contributions to human life, there were contrasting results, especially between the more negative opinions of women in El Salvador and the more positive opinions of those in Kamloops. Regarding age related opinions, which were generally weaker than those related to gender, Salvadorans showed a stronger generational gap, possibly because the near extinction of the big cats elicited more sympathy and regrets from the younger generation. In Canada the generational difference was stronger for bears, possibly due to their greater visibility and more variable behaviour. Relating the results of this literature review to the points listed as the key questions and hypotheses (the importance of the cougar's physical size and behaviour relative to the those of the jaguar and the bears), the evidence points to lack of specific knowledge of the differences between the cougar and the jaguar, and greater knowledge of bear behaviour, as key issues, these being superimposed on fairly strong but variable gender based opinions and weaker age related opinions. The contradictions between the Canadian examples also show that individual life experiences may be at least as important as generalised attitudes, as the former may skew results. The conclusion, based on these wide ranging but limited examples must also hint that the cougar's reputation as a dangerous animal is possibly larger than reality, easily matching that of the jaguar and the brown bear, and subordinate only to the troublesome behaviour of bears. Further studies may focus on actual relative impacts of these carnivores (measuring attacks on people and other animals, threatening behaviour and impacts of people's movements) for a comprehensive evaluation of the accuracy of social perceptions.

REFERENCES

Almeida, T. 1990. *Hunting in the Mato Grosso and Bolivia*. Safari Press, Long Beach.

Andersone, Z. & Ozolins, J. 2004. Public perception of large carnivores in Latvia. *Ursus.* 15: 181–7.

Batterbury, S, Forsyth, T. and Thomson, K. 1997. Environmental transformations in developing countries: hybrid research and democratic policy. *The Geographical Journal* 163: 126–32.

Beier, P. 1991. Cougar attacks on humans in the United States and Canada. *Wildlife Society Bulletin.* 19: 403- 412.

Beier, P. 1993. Determining minimum habitat areas and habitat corridors for cougars. *Conservation Biology.* 7(1): 94-108.

Boulhosa, R. L. P., and Valdes. S. A. C. 1997. *Impact of big cats predation on domestic cattle in the Pantanal Matogrossense, Brazil.* 4th Research report: September/96-August/97. CENAP/Associação Pró-Carnívoros, São Paulo.

Bowman, J. L., Leopold, B. D., Vilella, F. J., Gill, D. A., & Jacobson, H. A. 2001. Attitudes of landowners toward American black bear compared between areas of high and low bear populations. *Ursus.* 12: 153-160.

Bowman, J. L., Leopold, B. D., Vilella, F. J., & Gill, D. A. 2004. A spatially explicit model, derived from demographic variables, to predict attitudes toward black bear restoration. *Journal of Wildlife Management.* 68(2): 223-232.

Brown, Gary 1993. *The Great Bear Almanac*. Lyons & Burford, New York.

Burnien, D. and Wilson, D. E. 2001. *Animal: The Definitive Visual Guide to the World's Wildlife*. Dorling Kindersley, New York City.

Campbell, M. 1998. Interactions *between biogeography and rural livelihoods in the coastal savanna of Ghana.* Unpublished PhD thesis Imperial College University of London.

Campbell, M. 2005a. Sacred groves for forest conservation in Ghana's coastal savannas? Assessing ecological and social dimensions. *Singapore Journal of Tropical Geography* 26:151–69.

Campbell, M. 2005b. The ecological and social context of mammal hunting in the coastal savanna of Ghana *Geoforum* 36: 667– 80.

Campbell, M. and Lancaster, B. 2010 Public attitudes towards black bears *Ursus americanus* and cougars *Puma concolor* on Vancouver Island. *Journal of Society and Animals* 181: 40–57.

Campbell, M. and Torres-Alvarado, M. 2011. Public perceptions of jaguars *Panthera onca*, pumas *Puma concolor* and coyotes *Canis latrans* in El Salvador. *Area* 43(3): 250–256.

Campbell, M. Forthcoming for 2012, *Journal of Society and Animals*. The relevance of age and gender for public attitudes to brown bears (*Ursus arctos*), black bears (*Ursus americanus*) and cougars (*Puma concolor*) in Kamloops, British Columbia.

Carroll, C., Noss, R.F. & Paquet, P. 2003. Carnivores as focal species for conservation planning in the Rocky Mountain Region. *Ecological Applications* 11: 961–980.

Central Intelligence Agency 2010. The World Factbook: Central America and Caribbean El Salvador (*https://www.cia.gov/library/publications/the-world-factbook/geos/es.html*). Accessed 8 August 2010.

Conforti, V.L. & Azevedo, F.C.C. 2003. Local perceptions of jaguars *Panthera onca* and pumas *Puma concolor* in the Iguacu National Park area south Brazil. *Biological Conservation*. 111: 215–21.

Crawshaw, P. G. & Quigley, H. B. 2002. Jaguar and puma feeding habits. In *El jaguar en el nuevos milenio*. Una evaluacion de su estado, deteccion de prioridades y recomendaciones para la conservaci ´on de los jaguars en America: 223–235. Medellýn, R. A., Chetkiewicz, C., Rabinowitz, A., Redford, K. H., Robinson, J. G., Sanderson, E. & Taber, A. (Eds). Mexico: D. F. Universidad Autonoma de Mexico/Wildlife Conservation Society.

Davis, H., Wellwood, D., & Ciarniello, L. M. 2001. *"Bear smart" community program: Background report*. British Columbia Ministry of Environment, Lands and Parks, Victoria.

Decker D. J., Brown, T. L. & Seimer, W. F. 2001. *Human dimensions of wildlife management in North America*. Wildlife Society, Bethesda.

Deurbrouck, J. 2007. *Stalked by a Mountain Lion: Fear, Fact and the Uncertain Future of Cougars in America*. Falcon Press, Guilford.

Dickson, B. G. & Beier, P. 2007. Quantifying the influence of topographic position on cougar (*Puma concolor*) movement in southern California, USA. *Journal of Zoology*. 271: 270–277.

Foerster, C. 1996. Researcher attacked by puma in Corcovado National Park, Costa Rica. *Vida Silverstre Neotropical* 5: 57 -58.

Fulton, D. C., Pate, J., & Manfredo, M. J. 1995. *Colorado residents attitudes towards trapping in Colorado*. Colorado State University, Fort Collins.

Guggisberg, C.A.W. 1975. *Wild Cats of the World*. Taplinger Publishing, New York.

Hayward M.W. & Somers M.J. 2009. *Reintroduction of top-order predators*. Wiley-Blackwell, Oxford.

Hecht, S.B., Kandel, S., Gomes, I., Cuellar, N. & Rosa, H. 2006. Globalization, forest resurgence, and environmental politics in El Salvador. *World Development* 34: 308–23.

Herbert, D. 1989. *The status and management of cougar in British Columbia*. Proceedings of the Mountain Lion Workshop, 3, 11-14.

Hoogesteijnal, R. & Hoogesteijnal, A. 2008. Conflicts between cattle ranching and large predators in Venezuela: could use of water buffalo facilitate felid conservation? *Oryx*. 42 (1):132-138.

Inskip, C. & Zimmermann, A. 2009. Human–felid conflict: a review of patterns and priorities worldwide. *Oryx* 43 18–34.

Johnsingha, A.J.T. & Negib, A. S. 2003. Status of tiger and leopard in Rajaji–Corbett Conservation Unit, northern India. *Biological Conservation* 111: 385–393.

Kellert, S. R. 1976. Perceptions of animals in American society. *Transactions of the North American Wildlife and Natural Resources Conference*, 41: 533-545.

Kellert, S. R. 1989. Human-animal interactions: A review of American attitudes to wild and domestic animals in the 20th century. In A. Rowan (Ed). *Animals and People Sharing the World* (pp. 137-176). Hanover, NH: University of New England Press.

Kellert, S. R. 1994. Public attitudes towards bears and their conservation. *International conference on Bear Research and Management.* 9: 43-50.

Kellert, S. R., Black, M., Rush, C. R., & Bath, A. J. 1996. Human culture and large carnivore conservation in North America. *Conservation Biology*. 10: 977-990.

Iriarte, J. A., W. L. Franklin, W. E. Johnson, & K. H. Redford. 1990. Biogeographic variation of food habits and body size of the America puma. *Oecologia* 85: 185-190.

Johnsingh, A.J.T., 1983. Large mammalian prey-predators in Bandipur. *Journal of Bombay Natural History Society* 80: 1–57.

Jorgensen, J.P. & Redford, K.H. 1993. Humans and big cats as predators in the Neotropics. pp 367-390 in M.G.N Dunstone and M.L. Gorman, editors. *Mammals as predators. Series Symposia of the Zoological Society*. Oxford University Press, Oxford.

Karanth, K. U. & Sunquist, M. E. 1995. Prey selection by tiger, leopard, and dhole in tropical forests. *Journal of Animal Ecology* 64: 439–450.

Kleiven, J., Bjerke, T., & Kaltenborn, B. P. 2004. Factors influencing the social acceptability of large carnivore behaviours. *Biodiversity and Conservation*. 13: 1647-1658.

Maclennan, S. D., Groom, R. J., Macdonald, D. W. & Frank, L. G. 2009. Evaluation of a compensation scheme to bring about pastoralist tolerance of lions. *Biological Conservation* 142: 2419.

Maehr, D. S., Noss, R. F. & Larkin, J. L. 2001. *Large mammal restoration*. Island Press, Washington DC.

Maehr, D.S., Land, E.D., Shindle, D.B., Bass, O.L. & Hoctor, T.S. 2002. Florida panther dispersal and conservation. *Biological Conservation*. 106, 187–197.

McKee, D. 2003. Cougar attacks on humans: A case report. *Wilderness and Environmental Medicine*. 14:169–173.

Monroy-Vilchis, O., Rodríguez-Soto, C., Zarco-González, M. & Urios, V. 2009. Cougar and jaguar habitat use and activity patterns in central Mexico *Animal Biology* 59: 145–157.

Morzillo, A. T., Mertig, A. G., Garner, N., & Liu, J. 2007. Resident attitudes toward black bears and population recovery in East Texas. *Human Dimensions of Wildlife*. 12: 417–428.

Morzillo, A. T., Mertig, A. G., Hollister, J. W., Garner, N., & Liu, J. 2010. Socioeconomic factors affecting local support for black bear recovery strategies. *Environmental Management*. 45: 1299–1311.

Nowell, K. and Jackson, P. 2006. *Wild Cats. Status Survey and Conservation Action Plan*. IUCN/SSC Cat Specialist Group. IUCN, Gland.

Nuñez, R., Miller, B. & Lindzey, F. 2002). Food habits of jaguars and pumas in Jalisco, Mexico. *Journal of Zoology(London)*. 252: 373–379.

Nuñez, R., Miller, B. & Lindzey, F. 2002. Ecology of jaguars in the Chamela–Cuixmala Biosphere reserve, Jalisco, Mexico. In *El jaguar en el nuevo milenio*. Una evaluacion de su estado, deteccion de prioridades y recomendaciones para la conservacion de los jaguars en America: 107–126. Medellin, R. A., Chetkiewicz, C., Rabinowitz, A., Redford, K. H., Robinson, J. G., Sanderson, E. & Taber, A. (Eds). Mexico, D. F.: Universidad Nacional Autònoma de Mexico/Wildlife Conservation Society.

Oliveira, T. G. de. 1994. *Neotropical cats: ecology and conservation*. EDUFMA, São Luís,

Oliveira, T. G. de 2002. Comparative feeding ecology of Jaguar and Puma in the Neotropics. Ecología comparativa de la alimentación del jaguar y del puma en el neotrópico. Pp. 265-288 in *El jaguar en el nuevo milenio*. R. A. Medellín, C. Equihua, C. L. Chetkiewicz, P. G. Crawshaw Jr., A. Rabinowitz, K. H. Redford, J. G. Robinson, E. W. Sanderson, A. Taber, eds. Fondo de Cultura Económica/Universidad Nacional Autónoma de México/Wildlife Conservation Society, México, D.F. 2002.

Palmeira F.B.L, Crawshaw, P.G., Haddad, C.M., Ferrazd, K.M.P.M.B and Verdadee, L.M. 2008. Cattle depredation by puma *Puma concolor* and jaguar *Panthera onca* in central-western Brazil. *Biological Conservation*. 141: 118–25.

Pelton, M., Scott, C.. & Burghardt, G. M. 1976. Attitudes and opinions of persons experiencing property damage and/or injury by black bears in the Great Smoky Mountains National Park. *International Conference of Bear Research and Management*. 3: 157-167.

Perry, R. 1970. *The World of the Jaguar*. Taplinger Publishing. New York.

Polisar, J. D. 2000. *Jaguars, Pumas, their Prey Base, and Cattle Ranching: Ecological Perspectives of a Management Issue*. PhD dissertation, University of Florida.

Polisara, J. Maxit, I, Scognamillo, D., Farrell, L., Sunquist, M.E., Eisenberg, J.F. 2003. Jaguars, pumas, their prey base, and cattle ranching: ecological interpretations of a management problem. *Biological Conservation*. 109 (2): 297–310.

Sáenz, J. C. and Carrillo, E., 2002. Jaguares depredadores de ganado en Costa Rica: ¿un 982 problema sin solución? In *El Jaguar en El Nuevo Milenio*. ed. R. Medellin, C. Equihua, 983 C. Chetkiewicz, P. Crawshaw, A. Rabinowitz, K. F. Redford, J. Robinson, E. Sanderson 984 and A. Taber, pp. 127-138. Fondo de cultura económica, Universidad Nacional 985 Autónoma de México, Wildlife Conservation Society, México.

Schwartz, C.C., Swenson, J.E. & Miller, S.D. 2003. Large carnivores, moose and humans: a changing paradigm for the 21st century. *Alces* 39: 41-63.

Scognamillo, D., Maxit, I.E., Sunquist, M. and Polisar, J. 2003. Coexistence of jaguar (*Panthera onca*) and puma (*Puma concolor*) in a mosaic landscape in the Venezuelan Llanos. *Journal of Zoology (London)*. 259:269–279.

Soule′, M.E. & Terborgh, J. 1999. *Continental conservation*. Island Press, Covelo.

Taber, A.B., Novaro, A.J., Neris, N., & Colman, F.H., 1997. The food habits of sympatric jaguar and puma in the Paraguayan Chaco. *Biotropica* 29: 204–213.

Teel, T. L., Krannich, R. S., & Schmidt, R. H. 2002. Utah stakeholders' attitudes toward selected cougar and black bear management practices. *Wildlife Society Bulletin*. 30: 2-15.

Thornton, C. T., & Quinn, M. S. 2009. Coexisting with cougars: public perceptions, attitudes, and awareness of cougars on the urban-rural fringe of Calgary, Alberta, Canada. *Human–Wildlife Conflicts*. 3(2): 282–295.

Treves A and Karanth K U 2003. Human–carnivore conflict and perspectives on carnivore management worldwide. *Conservation Biology*. 176: 1491–1499.

Turner, J. W., Wolfe, M. L., & Kirkpatrick, J. F. 1990. Seasonal mountain lion predation on a feral horse population. *Canadian Journal of Zoology*. 70: 929-934.

Weber, W. and Rabinowitz, A.R. 1996. A global perspective on large carnivore conservation. *Conservation Biology*. 104: 1046–1054.

Williamson, D. F. 2002. *Status, management, and trade of the American black bear (Ursus americanus) in North America*. World Wildlife Fund, Washington DC.

Yàñez, J. L., Cardenas, J. C., Gezelle, P. and Jaksic, F. M. 1986. Food habits of the southernmost mountain lions (*Felis concolor*) in South America: natural versus livestocked ranges. *Journal of Mammalogy* 67: 604-606.

In: Advances in Zoology Research. Volume 6
Editor: Owen P. Jenkins

ISBN: 978-1-61122-061-2

Chapter 6

PATTERN FORMATION AND DIVERSITY IN BUTTERFLY WINGS: EXPERIMENTS AND MODELS

Toshio Sekimura
Department of Biological Chemistry, Graduate School of Bioscience and Biotechnology, Chubu University, Kasugai, Aichi, Japan

ABSTRACT

Butterfly wings are covered with thousands of monochromatic scale cells, which form regularly arranged patterns, that is, parallel rows of scale cells in the proximal-distal direction of the wing. Color pattern in butterfly wings is a kind of mosaic pattern of dozens to hundreds of finely-tiled monochromatic scale cells. The wing consists of two cell layers, a surface layer and a back layer. The two layers are completely separated by the middle tissues, and developmental phenomena in each layer occur independently of the other. Color pattern formation occurs in a single layer of the wing disc and it is essentially two-dimensional in space without cell movement. Therefore, the problem of color pattern formation could be solved by analyzing the underlying mechanism of how colored scales are settled at specific positions in a two-dimensional wing surface.

In this article, I review the progress of experiment-based mathematical modeling for color pattern formation and diversification in butterfly wings. First, I describe fundamental features of the wing color pattern, and then go into discussions on guiding principles or governing rules to derive from the so-called ground plan to diverse wing color patterns. In order to make the problem clear, I take up some specific butterfly species as model systems, of which color pattern and its diversity have been well investigated with both experiment and mathematical analysis. Finally, I would like to stress the importance of integration of different aspects such as gene and environment for understanding of the diversity in wing color patterns.

1. Introduction

Butterfly wing color patterns are one of the most colorful and spectacular examples of pattern in biology. The color patterns have evolved as mechanisms for visual communication such as camouflage, sexual signaling, warning coloration, and mimicry. Biologists have been interested in the developmental origin and evolutionary diversification of these patterns for more than a century. Comparative morphological studies revealed that a great unity underlies the diverse patterns of lepidopteran wings. Most butterflies have color patterns composed of discrete pattern elements whose homology can be traced across genera and family. The system of homologies among pattern elements is now called the nymphalid ground plan (Schwanwitsch, 1924; Süffert, 1927; Nijhout, 1991, see also section 8.1). Experimental work revealed that some of the pattern elements develop around discrete organizing centers, from which some signals affecting the spatial pattern of pigment synthesis are sent out in their vicinity. Recent experimental works on the developmental genetics of eyespot patterns have found out several genes related to the development of the eyespot organizing centers (e.g., Brunetti et al., 2001).

In parallel to the experimental work, several investigators have attempted to build experiment-based mathematical models for color pattern formation in butterfly wings (e.g., Murray, 1981; Bard & French, 1984; Sekimura et al., 2000 for global color patterns, and Nijhout, 1990 for eyespot pattern). In following sections, I review the progress of both experimental and mathematical researches on butterfly wing pattern and diversity.

2. Development and Characteristics of Butterfly Wing

Lepidopteran wings are composed of two monolayers of epithelial cells, which are separated by an extracellular space. Immediately after pupation, the epithelial cells of the wing are not differentiated and they are morphologically homogeneous. About one to three days after pupation (the timing depends on the insect and the temperature at which it develops), two cell types can be readily distinguished. The smaller cells are generalized epithelial cells (GECs) of the wing, and the larger cells are scale precursor cells (SPCs) that differentiate from GECs at the inception of adult development. SPCs are arranged in space such that they are separated from each other surrounded by GECs. Within a few hours of differentiation of the isotropically arranged SPCs, these cells become polarized along the proximodistal axis of the wing and begin to align into rows parallel to the anteroposterior axis of the wing. This row formation continues until a stable spatial periodicity of rows is established. These parallel rows of SPCs that are established at the beginning of adult development maintain their arrangement throughout adult development and represent the same rows of scales that appear on the surface of the adult wing (Nijhout, 1980; Nardi & Magee-Adams, 1986; Yoshida & Aoki, 1989). As a result, we could see thousands of scale cells in a highly ordered and intricate pattern on each surface of the adult wing.

3. Two Types of Patterns – Color Pattern and Cellular Pattern

As noted in the previous section, thousands of monochromatic scale cells cover the wing in a highly ordered cellular pattern, that is, parallel rows of scale cells in the proximal-distal direction of the wing. The parallel rows can be easily seen through a simple magnifying glass. There exist two different kinds of patterns, color pattern and cellular pattern of scale cells (Figure 1). The colors on butterfly wings are due to the colors of regularly-arranged monochromatic scale cells. The overall color pattern is constructed as a fine-tiled mosaic of overlapping, monochromatic scales, and it is characteristic of each lepidopteran species. The arrangement pattern of scale cells, that is, cellular arrangement pattern occurs in the early stages of pupation, while pigment syntheses related to color patterns occur in the last stage of pupation after completion of cell rearrangement. It is known that the formation of the color pattern is independent of the arrangement pattern formation of scale cells. The timescales on which these two kinds of patterns are generated are different from each other (Nijhout,1991). Color patterns have attracted the interest of taxonomists and developmental biologists and have also provided some of the most dramatic examples of morphological diversity. The color of scale cells is mainly due to chemical pigments, or it is sometimes due to the structural color that comes from diffraction of light in the physically fine-structured scale. On the other hand, cellular arrangement patterns of scale cells do not show species-specific patterns, but they are, in general, common to all lepidopteran wings. The arrangement of scale cells has a number of remarkable characteristics. Firstly, SPCs in the developing adult wings rearrange to form nearly parallel rows along the anteroposterior axis of the wing that are maintained throughout and after adult development. Secondly, these rows are arranged at regular spatial intervals along the proximodistal axis. Thirdly, these rows are continuous across the veins in the wing, and they are also continuous around the upper and lower monolayers of the wing.

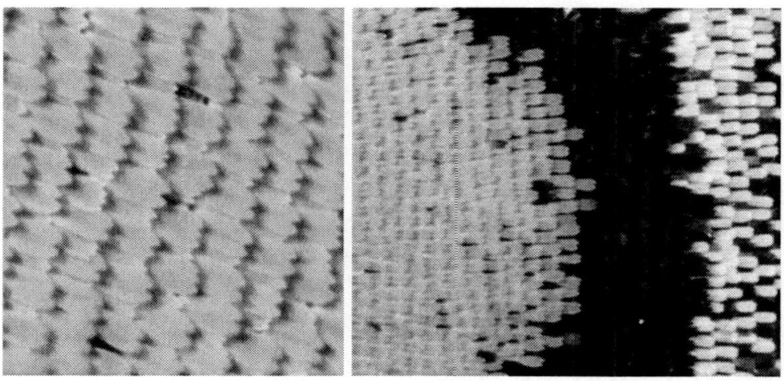

Figure 1. Cellular pattern and color pattern. (Left) Cellular pattern: parallel rows of monochromatic scale cells in the proximal-distal direction of the wing. (Right) Color pattern: mosaic pattern of dozens to hundreds of finely-tiled monochromatic scale cells, which is characteristic of each lepidopteran species.

I first describe cellular pattern formation of scale cells as the basis of all patterns in butterfly wings and then go to color pattern formation with mathematical models.

4. Cellular Pattern Formation – Experiments and Model

I first summarize experimental evidences (i) ~ (v) to construct a mathematical model for cellular pattern formation in the lepidopteran wing, and then present a mathematical model and computer simulation results (Sekimura et al., 1998,1999).

4.1. Experimental Evidences

i. Cell rearrangement to form parallel rows occurs in a monolayer

There are two monolayers (the upper and lower ones) of epithelial cells in the lepidopteran wing. During rearrangement of epithelial cells in the wing, these two monolayers are separated by an extracellular space. Cell rearrangement in the developing adult wing has been assumed to occur within each monolayer without influence from the other monolayer (Nardi, 1992, 1994). These findings suggest that the process of scale row formation in the lepidopteran wing does not involve complex three-dimensional interactions among epithelial cells, but instead involves simpler two-dimensional interactions.

ii. Lateral inhibition probably forms the uniform pattern of SPCs

Before epithelial cells rearrange their spatial positions in the developing adult wing, SPCs differentiate from GECs. SPCs are distributed in an uniform hexagonal pattern, in which they do not come in contact with each other. The initial isotropic spacing of SPCs is strongly suggestive of patterns generated by a process referred to as lateral inhibition. The inhibitory signal emanates from the scale precursor cell.

iii. Short-range interaction mediated by differential cell adhesion

A few kinds of surface proteins have been found in the developing adult wing monolayer during rearrangement of epithelial cells. The expression of these proteins changes dramatically in space and time during morphogenesis of the wing pattern. SPCs show a different pattern of protein expression from GECs during cell rearrangement. These proteins are assumed to be cell adhesion molecules (CAMs), which mediate differential cell adhesion between different cell types (short-range interaction) to drive cell movement.

iv. Origin-dependent cell adhesion

Grafting experiments within the pupal wing monolayer have been done to test differences in affinity properties of epithelial cells along the proximodistal axis of the wing. Those experiments have revealed that the greater the distance separating host and graft cell populations along the proximodistal axis, the more circular and constricted the interface between graft and host cells. This result shows that affinity properties of cells depend on their positions, or more precisely, the distance between their original positions along the proximodistal axis of the wing.

v. Long-range interaction mediated by basal processes

During rearrangement of epithelial cells, both SPCs and GECs extend basal processes. As the alignment of SPCs into rows proceeds, extension of processes from the basal surfaces of the epithelial cells simultaneously occurs. These processes can extend for distances of several cell diameters and can establish contacts not only with adjacent cells but with cells that are four or five cell diameters away. This long-range interaction is mediated by basal filopodia which extend over several cell diameters in many directions (Nardi & Magee-Adams, 1986).

4.2. Mathematical Model for Cellular Arrangement Pattern of SPCs

Based on observation on cellular pattern formation, we begin by presenting a general model for aggregation in which cells move up gradients of adhesivity. Since cells can respond directly to non-adjacent neighbors, we use integrals to represent the local average adhesivity to which a cell responds. In this model we focus on only one cell type (SPCs) and assume that cells interact with each other according to the distance between their original locations (as well as the distance between their current locations).

Let $n(\vec{x},a,t)$ denote the cell density at position $\vec{x}=(x,y)$ at time t for the cells of a given adhesivity a that originate from a position that is a distance a away from the body axis (the basal part of the wing). We assume that cell movement is due to two processes; diffusion, and advection (directed movement) in response to gradients of adhesivity. Due to the evidence for long-range interactions, we consider a cell to respond to gradients in a spatially averaged adhesivity. As an evolution equation for cell density in space, we write

$$n_t = D\nabla^2 n - \nabla\cdot(n\vec{c}) \qquad (1)$$

where D is the diffusion coefficient. The advection velocity, $\vec{c}$, is given by

$$\vec{c} = C\nabla\left[\iiint n(\vec{x}-\vec{y},a-s)\; w(\vec{y},s)\,ds\,dy_1 dy_2\right]$$

where $\vec{x}=(x,y)$ and $\vec{y}=(y_1,y_2)$ are position variables, and C is a positive constant. The integral represents the spatially averaged adhesivity. The degree of adhesivity as a function of distance, $\vec{y}$, and adhesivity distance (distance in adhesivity space), s, are incorporated in the kernel $w(\vec{y},s)$. For simplicity, we assume that this is separable (i.e., that the effects of distance in physical and adhesivity space are independent of each other). We therefore write

$$w(\vec{y},s) = g(\vec{y})\; h(s)$$

We assume that g displays rotational symmetry in the two spatial dimensions, and h is symmetric in the adhesivity difference. Notice the threshold in adhesivity marking a transition from attraction to repulsion. Also, the spatial kernel is such that very short range attraction is weaker than middle distance attraction. This reflects the fact that scale cells appear not to come into contact with one another during reorganization.

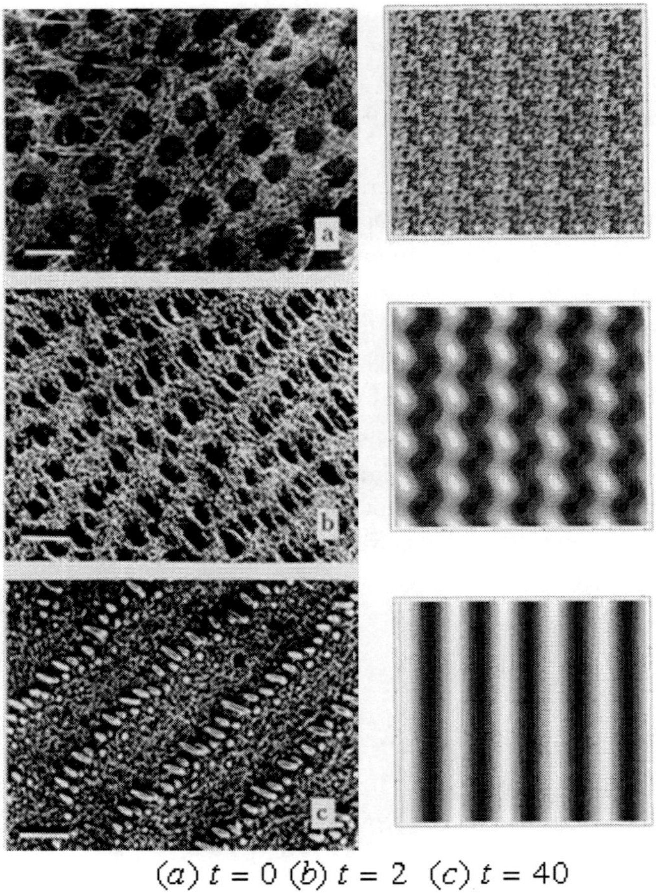

Figure 2. Experiments and numerical simulations. (Left) Parallel row formation of SPCs on the moth *Mandusa* wing disc (Nardi & Magee-Adams, 1986). Bar in the figure= 50 μm. (Right) Numerical simulation results on the row formation by the mathematical model Eq. (2) (Sekimura et al.,1998). The lighter color represents high cell density of SPCs, while the darker color represents low cell density. In numerical simulations, a random cell distribution is assumed as the initial condition, and periodic boundary conditions have been imposed on the boundary.

Below, we assume that $|\vec{y}|$ and $|s|$ are both much smaller than one. Biologically, this corresponds to the assumption that the adhesive interaction takes place over short distances in comparision to the size of the domain and that changes in the magnitude of adhesivity are very small. Using the assumption that $|\vec{y}| << 1, |s| << 1$, we can simplify the integral represention of the advection velocity, $\vec{c}$, by Taylor expanding, and finally we could have a following partial differential equation:

$$n_t = D\nabla^2 n \; - \; C\nabla \cdot \left[n\nabla(n + \gamma\nabla^2 n + \beta n_{aa}) \right] + \; O(s^4 + |\vec{y}|^4)$$

where,

$$\gamma = \frac{1}{2}\int |\vec{y}|^2 g(\vec{y})\, dy_1\, dy_2\,,\ \beta = \frac{1}{2}\int s^2 h(s)\, ds$$

The above equation includes parameters γ, β, which represent two different kinds of adhesivity effects, physical distance effect and adhesive distance effect, respectively. We further assume that $n_{aa} = E n_{xx}$ (E: constant) and we neglect the last term. The equation for cellular pattern (parallel row) formation of SPCs with origin-dependent adhesivity is,

$$n_t = D\nabla^2 n \; - \; C\nabla\cdot\left[n\nabla(n + \gamma\nabla^2 n + \beta^* n_{xx})\right] \qquad (2)$$

where, $\beta^* = E\beta\ (\leq 0)$. Here we assume an inequality $\gamma + \beta^* \geq 0$, showing that due to the effect of origin-dependent adhesivity, SPCs receive a kind of reduction in diffusive movement in the $x-$ direction (the proximal-distal direction of the wing). According to our non-linear analysis of the equation, we have found following three main results. The first result is that due to the effect of origin-dependent adhesivity, the region in the parameter space of D, C, γ to generate stable parallel rows of SPCs is largely extended. The second result is more important, that is, due to the effect of origin-dependent adhesivity, the direction of parallel rows is correctly predicted (Figure 2). This means that due to the effect, scale rows parallel to the body axis seen in real butterflies could be correctly reproduced. The third result is that if the effect is sufficiently strong, only parallel rows of SPCs could be generated without spotted patterns.

5. Color Pattern Formation

5.1. General Features of Color Pattern Formation – Color Pattern and Pre-Pattern

Butterfly color patterns are two-dimensional patterns of pigment synthesis. A fraction of the epidermal cells on each wing surface differentiate into scale cells. These cells send out a large flat appendage, the scale, into which pigments are secreted. With few exceptions, each scale cell synthesizes a single type of pigment. The overall color pattern is thus constructed as a fine-tiled mosaic of colored scales. One of the problems of pattern formation is how a particular scale cell is induced to synthesize the right pigment for its particular location on the wing.

Recently, antibody fluorescent techniques revealed that the pattern formation has started at least in the late larval stage of development. The butterfly wing originates from the wing disc in the larva, which appears first in the larval stage and continues to develop during the pupal stage. The wing disc is transparent and colorless throughout the larval and early pupal stages of development. However, antibody fluorescent techniques for the corresponding genes have revealed existence of the so-called "pre-pattern" on the wing disc, which overlaps completely with the color pattern of the adult butterfly wing (e.g., Brunetti et al., 2001). It is generally known that the pre-pattern on the wing disc is the original feature of the color pattern. Pattern determination in butterflies has been considered to occur in two stages. The first stage consists of the specification of organizing or signaling centers on the wing surface. In species where this has been studied, this process occurs during the early to mid portions of the larval stadium in the still growing wing imaginal disc (Nijhout, 1991; Carroll et al., 1994). The second stage consist of the definition of the boundaries of the future patterned synthesis of each of the pigments that will make up the color pattern, which are established by signals generated by the organizing centers. This second stage results in an invisible spatial pre-pattern of cell commitment to a particular pigment synthetic pathway. Actual pigment synthesis does not begin until the end of the pupal stage, a few days before emergence of the adult butterfly. We therefore assume that the problem of color pattern formation is essentially that of pre-pattern formation in the wing disc, but not simple color setting by pigmentation in the latest pupal stage. Since the wing disc grows in size and shape, we in deed have to understand the time-course of development of pre-pattern formation during the time period of larval and pupal stages.

5.2. Two Types of Color Patterns – Local Color Pattern and Global Color Pattern

Local color pattern means localized pattern in a restricted area of the entire wing surface such as the eyespot pattern. Global color pattern, on the other hand, means the whole pattern covering whole dorsal and ventral wing monolayers and it is sometimes used for identification of butterfly species (Figure 3).

Figure 3. Local color pattern and global color pattern. (Left) Local eyespot patterns with colored concentric rings on the forewing of *Micalesis gotama*. (Right) Global color patterns on fore- and hindwings of *Papilio xuthus*.

Global color pattern includes local color patterns as parts of all pattern elements composing the global wing pattern. We should have in mind that research on local color patterns is not independent of research on the global color pattern. Recent research on the genetics of color patterns has been mostly concentrated on the local color pattern. On the other hand, genetic research on the global pattern has not been so far advanced. Recently, however, great progresses have been made using mimetic butterflies such as *Papilio dardanus*, *Heliconius* butterflies, and *Papilio polytes*.

However, the problem of color pattern fomation in wings is still not sufficiently solved through both experiment and theory. A few mathematical models based on experimental data have been proposed to account for specific features of the pattern and I review these models briefly with experimental data in following sections.

6. Local Color Pattern Formation – Experiments and Models

6.1. Genetic Research on Eyespot Pattern Formation

Among local color patterns, the eyespot pattern has been studied most at both molecular and genetic levels. Carroll et al. (1994) found eye-spot foci in the wing disc of a species *Precis coenia* by using antibody fluorescent techniques for the gene *Distal-less*. The positions of the foci on the wing disc were found to correspond completely to eyespot centers on the adult butterfly (Brakefield et al.,1996). Brunetti et al. (2001) found that two different transcription factors, *spalt* and *engrailed* are expressed in the wing disc of the African species *Bicyclus anynana.* The expression patterns of concentric rings, that is, eyespot pre-patterns correspond completely to the eyespot patterns on the adult butterfly. Some genes associated with pigmentation have also been investigated and identified so far. Current researching trends for eyespot pattern formation are in the direction of detailed genetic analyses, which have taken into account newly founded gene regulatory networks. Based on these findings, the mechanism of eyespot formation has been assumed to consist of following three developmental stages,

(i) Determination of the eyespot focus, or signaling center in the wing disc, from which some signaling chemicals, or morphogens originate. The focus point on the disc is determined by complex genetic regulatory networks.
(ii) Spreading out of morphogens into the surroundings of the focus point through diffusion, and activation of corresponding genes (e.g., *Distal-less, engrailed*) to establish concentric eyespot pre-patterns on the disc.
(iii) Activation of pigmentation genes (e.g., *DDC, GTP-CH1, cinebar*) depending on pre-patterns, and then, color differentiation of scale cells in the adult wing.

6.2. Genetic Research for Local Color Patterns with Respect to Evolution

Reed & Serfas (2004) analyzed changes in the *Notch/Distal-less* intervenous expression patterns in wing discs of 8 different butterfly species such as *Vanessa cardui, Precis coenia and Bicyclus anynana*, with respect to evolution of pattern elements. They concluded that the combination of the *Notch/Distal-less* expression patterns is a good index or tool for understanding the evolutionary pathway of butterfly species. Monteiro et al. (2006) did a comparative homology research on saturniid moth and nymphalid butterfly eyespots with concentric rings of color, in which two transcription factors, *Distal-less* and *engrailed*, are found to express in the center of the future pattern. They also found that the ligand Wingless, and an activated signal transducer, a phosphorylated Smad protein, are expressed, but neither these proteins nor the previous two proteins are found in pierid spot centers that consist of a single patch of color. Both butterfly wing patterns, however, express a third transcription factor, *Spalt*, a portion of whose expression domain maps to the black-colored scales on the adult wing.

On the other hand, there are few mathematical models for local color pattern formation except for eyespot formation. Next, I would like to review mathematical models for eyespot formation.

6.3. Mathematical Models for Eyespot Pattern Formation

6.3.1. Activator–Inhibitor Model – A Model for Determination of the Eyespot Focus

Nijhout (1990) found that computer simulations by using a mathematical model could reproduce a focal point on the wing cell as a stable point-like solution of the model equation. The wing cell, in this case, means a localized area composed of scale cells between two adjacent veins near the wing margin, but not a biological single cell.

Eyespots develop around small groups of cells that act as signaling or inducing sources, which are sometimes called "organizing or signaling centers". Nijhout (1990) showed that point-like patterns can be produced in the exact locations of the organizing centers by an activator–inhibitor reaction-diffusion mechanism that assumes the existence of two chemicals: an activator, $a(\vec{x},t)$ at space $\vec{x}$ and time t, whose synthesis is subject to positive feedback (autocatalysis), and a inhibitor, $h(\vec{x},t)$, which suppresses autocatalysis around an area of activator production. The system is described by the following equations:

$$\frac{\partial a}{\partial t} = \frac{ca^2}{h} - k_1 a + D_a \nabla^2 a$$

$$\frac{\partial h}{\partial t} = ca^2 - k_2 h + D_h \nabla^2 h$$

where parameters c, k_1, and k_2 are reaction constants; D_a and D_h are diffusion constants; and the equations were solved in a two–dimensional space. In numerical simulations, it is assumed that the wing veins act as fixed boundary conditions for the activator and as reflecting boundaries for the inhibitor. In biological terms this implies that the veins, or cells associated with the wing veins, are sources of one of the diffusing reactants. Variation in parameter values can displace the positions of these point-like patterns, and mimic the effect of certain mutations on the positions of the centers of eyespots (e.g., Brakefield, 1998).

The mathematical model also predicted a rather complex time-evolution of the spatial pattern of activator concentration, before it settles to point-like pattern at the stable steady state (Nijhout, 1990). Several years later, Carroll et al. (1994) demonstrated that the gene

Distal-less is expressed uniquely in the cells of the organizing centers of eyespots. Moreover, the spatial pattern of expression of *Distal-less* undergoes exactly the same complex changes as those predicted by the mathematical model (Carroll et al., 1994; Nijhout, 1994).

6.3.2. Other Mathematical Models for Eyespot Pattern Formation

Current researching trends for eyespot pattern formation are in the direction of detailed genetic analyses, which have taken into account newly founded gene regulatory networks. Mathematical modeling trends are also in the same direction. Evans & Marcus (2006) presented a mathematical model for determination of the eyespot focus with gene regulatory networks.

As noted in section 6.1, the mechanism of eyespot pattern formation consists of three ((i), (ii), and (iii) stages in section 6.1) developmental stages. After determination of focus points, that is, after the first stage, morphogens spread out from the focus point into the surroundings through diffusion, and then, corresponding genes are activated to form pre-patterns, and finally, color differentiation occurs in scale cells just one or two days before emergence of adult butterflies. Dilão & Sainhas (2004) presented a reaction-diffusion model for eyespot pattern formation, which includes two diffusive morphgens and three non-diffusive pigment precursors. The model has focused on the process of the second and third stages of development for eyespot formation. Assuming that the focus point is given a priori in a specific region of the wing, the model has simulated successfully two concentric dark- and light rings of the pattern, the structural organization of eyespots, their phenotypic plasticity and seasonal variability.

7. Global Color Pattern Formation: Experiments and Models

7.1. Genetic Research on Global Color Patterns

Recent research on the genetics of color patterns has been mostly concentrated on the local color pattern and genes such as *Distal-less* and *engrailed* controlling the eyespot pre-pattern have already been identified (e.g., Brunetti et al., 2001). On the other hand, genetic research on the global pattern has not been advanced so far. Recently, however, great progresses have been made by using mimetic butterflies such as *Papilio dardanus*, *Heliconius* butterflies, and *Papilio polytes*. Among them, I select here two mimetic butterfly species, *Papilio dardanus* and *Heliconius erato*, which have been of special interest to biologists for over a century because of the adaptive variation in their wing patterns.

7.1.1. Papilio Dardanus

Papilio dardanus is a species of swallowtail butterfly widely distributed across sub-Saharan Africa. *P. dardanus* is well known for the spectacular phenotypic polymorphism in females. The females of different geographic races have evolved more than a dozen different wing color patterns and have come to mimic different species of unpalatable butterflies and moths in their specific regions (Batesian mimicry). The males, on the other hand, are monomorphic and strikingly different from the females, exhibiting a characteristic yellow and black color pattern and tailed hind wing (see Figure 4).

The breeding work of Clarke & Sheppard (1959, 1960) showed that the female-limited polymorphism is controlled by a single genetic locus, termed *H*, with at least 10 distinct alleles. Recently, Clark et al. (2008) demonstrated that the transcription factor *invected*, and its closely linked paralogue *engrailed*, are candidates for the mimicry locus *H*. In addition to genetic studies, other work using phylogenetics (Vane-Wright et al.,1999), morphometrics (Nijhout, 2003) and mathematical modeling (Sekimura et al., 2000) has been done to clarify the function and genetic architecture of *H*.

Figure 4. Polymorphism in females and male of *P. dardanus*. Top left photo shows male pattern, and other eight photos show different female forms of a species *P. dardanus*.
(Courtesy of Mr. Steve Collins of African Butterfly Institute, Nairobi, Kenya)

7.1.2. Heliconius Butterfly

It is known that the species *Heliconius erato* is a mimetic butterfly with more than 20 different wing color pattern variants adapted to different regions across Central and South America (Müllerian mimicry). Recently, genetic analyses of *Heliconius* have started to unearth the characteristics of global wing color patterns. Reed & Gilbert (2004) studied the relationship between wing venation and *Distal-less* expression in *Heliconius* displaying a mutant phenotype with a severe vein deficiency, and found that major color pattern elements develop independently of wing venation. This finding lends support to previous theoretical models suggesting that proximal-distal pre-patterns can be sufficient for establishing complex whole-wing color patterns in butterflies (Sekimura et al., 2000). Reed et al. (2008) investigated expression of candidate genes potentially involved with a red/yellow forewing band polymorphism in *H. erato*. They found that a transcript *cinnabar* expression was associated with the forewing band regardless of pigment color, and another *vermilion* expression changed spatially over time in red-banded butterflies, but was not expressed at detectable levels in yellow-banded butterflies, suggesting that regulation of *vermilion* may be involved with the red/yellow polymorphism.

I next review some mathematical models for global color pattern formation in pecific moth and butterfly wings.

7.2. Mathematical Models for Global Color Pattern Formation

7.2.1. Diffusing-Morphogen-Gene-Activation Model (Murray, 1981)

A simple model for the development of the commonly observed crossbands of pigmentation shortly after pupation was proposed by Murray (1981). This model is based on a diffusing-morphogen-gene-activation system and extends the idea of a determination stream proposed by Kühn & von Engelhardt (1933), namely, that the anterior and posterior margins of the wing are sources from which emanates a wave of morphogen concentration. As the morphogen diffuses, the morphgen is assumed to be degradated via first-order kinetics. The governing equation for the morphogen concentration $S(\vec{x},t)$ at space $\vec{x}$ and time t is then

$$\frac{\partial S}{\partial t} = D\nabla^2 S - k S$$

where $D(cm^2 s^{-1})$ is the diffusion coefficient and $\kappa(s^{-1})$ the degradation rate constant. Boundary conditions for the morphogen at the wing edges are taken to be zero flux.

As the morphogen diffuses across the wing surface, the cells react in response to the local morphogen concentration, and a gene activated by S produces a product g. The kinetics of the gene product is assumed to exibit a biochemical switch behavior produced by a standard cubic-like form.

$$\frac{dg}{dt} = k_1 S + \frac{k_2 g^2}{k_4 + g^2} - k_3 g$$

where $k_1 \sim k_4$ are positive parameters. Here, the product g is produced linearly by the morphogen S , by the product in a nonlinear positive feedback way and linearly degradated proportional to itself. It is noteworthy to notice that g is a function of position through the morphogen $S(\vec{x},t)$.

Murray's model hypothesizes that the morphogen activates a gene product which, in turn, determines color pattern. The above model equations are solved either on a sector of a circle, representing a wing, or a rectangle, representing a wing cell with appropriate boundary conditions (see Murray, 1981, 2003). Murray showed that this simple model could exhibit a wide variety of observed patterns. For example, it exhibits patterns consistent with those observed after microcautery surgery. The theoretical results are consistent with the observation of Schwantwitsch (1924). Varying the scale and geometry of the model wing leads to more complicated patterns that are consistent with those observed on certain butterfly and moth wings, while considering the model on a wing cell and varying the source strengths also leads to commonly observed patterns.

7.2.2. Simple Diffusion Model (Bard & French, 1984)

A simple diffusion mechanism of a single morphogen was proposed for the formation of the wing pigmentation patterns of three species of butterflies by Bard & French (1984). Their model is based on three main assumptions. First, there may be morphgen sources in the foci of eyespots and morphogen sinks at some parts of the wing margin. Second, the morphgen has a finite half-life and diffuses simply and freely away from the sources throughout a wing of hexagonally packed cells. Third, that the overt pattern derives from cells interpreting the local morphogen concentration with respect to thresholds which determine scale colors. The final pattern follows lines of constant morphogen concentration and may, depending on the distribution of sources, comprise rings, curves, or bands.

In the model, the morphogen was assumed to move through the wing epitherium by simple, membrane-limited diffusion. The morphogen was also assumed to be lost exponentially with time either by simple decay or by a first-order biochemical degradation. Thus, the concentration of the morphogen $C(\vec{x},t)$ at space $\vec{x}$ and time t is given by the equation

$$\frac{\partial C}{\partial t} = \mu \nabla^2 C - k C$$

where μ is the morphogen diffusion constant and κ is the decay constant. Since relatively few parameters are required to specify the system, the diffusion constant μ, the decay constant κ and the threshold values are the most important. They chose the diffusion constant to be 5×10^{-7} as a biologically reasonable value. Values of the morphogen half life and thresholds were chosen by trial and error. With a biologically realistic diffusion constant and a morphogen half-life less than 6 hours, the model could generate many features of butterfly wing patterns as stable patterns of morphogen concentration, the compound spots of *Tenaris domitilla*, the large rings of *Diaethria marchalii* and the pattern of eyespots, rings and asymmetric bands of *Ragadia minoa*.

7.2.3. Reaction Diffusion Model (Sekimura et al., 2000)

A reaction diffusion model on a geometrically accurate wing domain was first presented for the formation of global pigmentation patterns in the butterfly wing of *Papilio dardanus* by Sekimura et al. (2000). As noted in section 7.1.1, the female wing patterns look very complicated in their appearance and at first glance it seems difficult to find an underlying logical relationship between them even within the single species. However, Nijhout (1991) proposed an idea that the black color pattern elements in the wing constitute the principle pattern elements, even though the background color attracts our attention most. The elements differ in size depending on the mimetic form and this can have dramatic effects on the overall appearance of the pattern. The problem is, then, largely simplified and the goal is to present a mechanism that can account for only the black pattern elements.

The mathematical model is based on the idea that a system of reacting and diffusing chemicals could evolve from an initially uniform spatial distribution to concentration profiles that vary spatially – the diffusion driven instability (Turing, 1952). The model equations is the non-dimensionalised reaction-diffusion system with Gierer-Meinhardt (1972) reaction kinetics,

$$\frac{\partial u}{\partial t} = \gamma\left(a - bu + \frac{u^2}{v(1 + ku^2)}\right) + \nabla^2 u \tag{3}$$

$$\frac{\partial v}{\partial t} = \gamma\left(u^2 - v\right) + d\,\nabla^2 v \tag{4}$$

where $u(\vec{x},t)$ and $v(\vec{x},t)$ are chemical concentrations at position $\vec{x}$ and time t; a, b, d, K and γ are positive parameters.

[Result 1] Mimetic and Non-Mimetic Patterns in Females

Mathematical and computational analyses were made first with nine forms of different wing color patterns, that is, four mimetic forms (f. *trophonius*, f. *cenea*, f. *planemoides*, and f. *hippocoonides*) (see Figure 5), four non-mimetic forms (f. *natalica*, f. *niobe*, f. *salaami*, and f. *leigh*) of the female, and the male of *P. dardanus* (Sekimura et al., 2000). The idea for the analysis is based on the so-called ground plan of *P. dardanus* (Nijhout,1991), which consist of three regions on the forewing and one region on the hindwing, where (i) the black pattern elements are major parts of the wing patterns, even though the background color attracts our attention most, and (ii) the black pattern elements can increase or decrease in size, depending on the phenotypic form of each female.

By mathematical analysis and computer simulations of the model equations: Eqs. (3) & (4), Sekimura et al. suggested that the global wing coloration is essentially due to underlying stripe-like patterns of some pigment inducing morphogen. The result is consistent with experimental results on the *Heliconius* butterfly (Reed & Gilbert, 2004). Furthermore, their results suggest that different patterns in females are similar to each other and they could be generated essentially by a fixed set of kinetic and diffusion parameter values in a reaction-diffusion system. This result could be also important from the genetic point of view, because it agrees with the finding that different forms of the females are controlled by a single genetic locus (Clarke & Sheppard, 1959, 1960).

The model does not need to assume supplementary morphogen sources and sinks in some parts of the wing to get patterns. Sekimura et al. also stressed the importance of some key factors to have realistic patterns in computer simulations such as parameter values for mode selection, threshold values which determine color, wing shape and boundary conditions.

Figure 5. Mimetic forms in females of *P. dardanus* and numerical simulation results. (Left) Photographs of mimetic female forms of *P. dardanus* (f. *trophonius* (top left), f. *cenea* (top right), f. *planemoides* (bottom left), and f. *hippocoonideas* (bottom right)), (Courtesy of Dr. Alfried Vogler of Natural History Museum, and Imperial College, UK). (Right) Numerical simulation results by the mathematical model (Eqs. (3) & (4)) of corresponding forms: f. *trophonius*, f. *cenea*, f. *planemoides*, and f. *hippocoonideas*, respectively.

[Result 2] Male and Male-Like Female Patterns

To our surprise, females exibit male-like forms with wing tails in some populations. Difference between male and male-like female patterns is, however, very clear when we look at both front wing patterns closely. Yellow-colored patterns of male front wing are very simple and circularly outlined on the distal margin. Male-like female patterns, on the other hand, are somewhat structured and characteristic on the anterior margin of yellow pattern, showing a black colored hollow from outside into the center of the wing, which usually specify the phenotypic indication of male-like female forms. The problem of difference between male and male-like female patterns has been solved mathematically as follows.

From mathematical point of view, the male patterns are essentially different from the male-like female patterns. It was shown that the male patterns are related to parameter values $(d, \ \gamma)$ of the equations to isolate the (1,0) mode on a square domain. On the other hand, the structured male-like female patterns are assumed to relate to parameters with the (3,0) mode which could generate other female patterns such as that of the form f. *hippocoonides* (Nijhout et al., 2003).

8. Factors of Diversification of Wing Color Patterns

8.1. Ground Plan, a Hypothetical Origin of Global Color Patterns

Owing to the pioneering work of Schwanwitsch (1924) and Süffert (1927) on the nymphalid ground plan, the global color patterns on the wings can be understood as a composite of a relatively small number of pattern elements. For example, (1) the symmetry system consists of color bands that run anterior to posterior across the wing; (2) the border ocelli system consists of a series of eyespots in the distal half of the wing; (3) the marginal bands are a pair of narrow bands near the wing margin; (4) the dependent patterns are venous stripes, that is, a color pattern of the outline of the wing veins; (5) the ripple patterns run proximal to distal of the wing in a manner similar to the ripples in windblown sand (Figure 6).

It is generally assumed that diverse global color patterns originate from the specific ground plan of each species through morphogenetic changes caused by some factors. In another word, species-specific color patterns develop because of the selective expression or suppression of individual pattern elements, and also the developmental regulation of the exact shape and pigmentation of each element.

I next describe major key factors that produce diverse global color patterns.

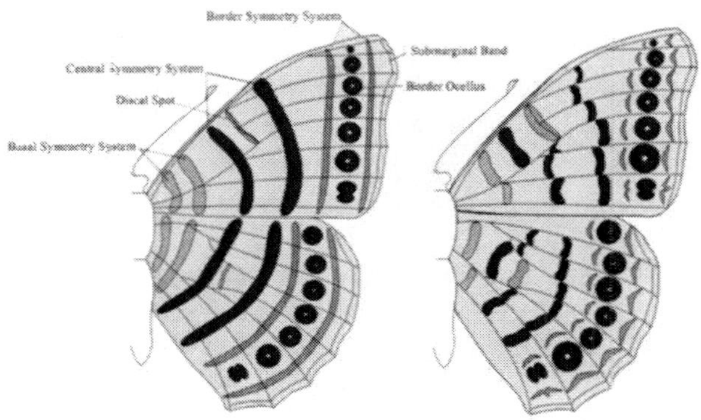

Figure 6. The Nimphalid ground plan according to Nijhout (1991). The ground plan consists of a system of pattern elements such as bands and spots that run from the anterior to the posterior margin of the wing. The pattern elements are classified and named as symmetry systems, submarginal bands, border ocelli, and discal spot.

8.2. Major Factors of Diversification of Global Color Patterns

In order to understand the diversity in global color patterns, we must know the vein system and vein positions in the wing, which play significant roles in generating the diversity. Precursors of wing veins, that is, pre-veins, appear first in the larval wing disc and develop throughout the larval and pupal stages until the appearance of adult butterfly. We focus our attentions on the relationship between the pre-veins and the pre-pattern formation in relation to the diversification of global wing patterns. Following factors (i)~(v) are known to have major effects to diversification of global wing patterns (see also Figure 7).

(i) *Disruption*

Disruption means that pre-veins divide pattern elements such as marginal bands running parallel to the wing margin into fragmental elements such as the border ocelli mentioned above. As an example, we can see in the species *Bicyclus anynana* a mutant without a wing vein showing that two adjacent eyes-pots merge into a single continuous pattern element (Brakefield et al., 1996).

(ii) *Dislocation*

Dislocation means that pre-patterns running parallel to the wing margin lose their continuity over pre-veins by a kind of sliding movement along the pre-veins.

For example, Koch & Nijhout (2002) found a naturally occuring veinless mutant of the swallowtail butterfly *Papilio xuthus* showing extremely aberrant patterns, where the loss of veins changes normal dislocated marginal bands into continuous bands running parallel to the wing margin.

(iii) *Venous and intervenous patterns*

The venous pattern appears precisely in the line of the wing veins, and intervenous pattern, by contrast, consists of a pigmented stripe that lies on the middle of two adjacent veins.

(iv) *Vein-independent patterns*

This pattern is completely different from the above vein-dependent ones, and is usually based on pattern elements such as the central symmetry system running in the central part of the wing far from the wing margin. For example, in *Heliconius* butterflies, the major pattern elements, that is, stripe-like patterns running through the middle of the wing, appear to be vein-independent.

(v) *Expansion, shrinking, and elimination of pattern elements*

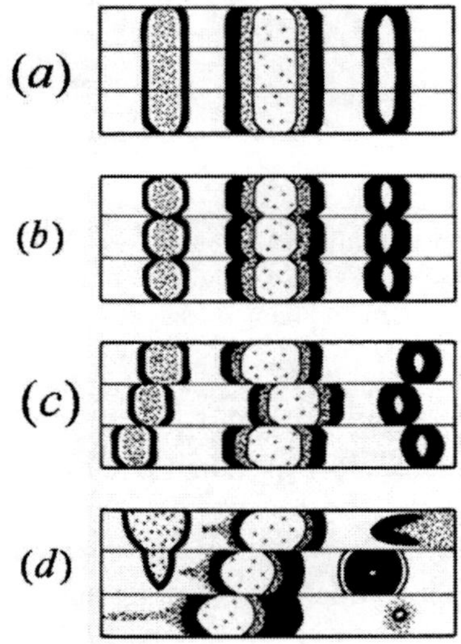

Figure 7. Major factors of diversification of color patterns. (a) Symmetry system of color bands that run anterior to posterior across the wing; (b) Disruption; (c) Dislocation; (c) Expansion, shrinking, elimination, and other deformation of pattern elements. (Courtesy of Prof. Fred Nijhout, Duke University)

8.3. Ecological Aspect: Factors Affecting Diversification of Global Color Patterns

Global color patterns are sometimes used for identification of butterfly species, because it is considered that diverse wing color patterns have been caused or derived from various types of adaptation to surrounding environments such as mimicry and behavior of females to choose food plants on which to lay eggs, and so on. Among ecological factors associated with evolutionary strategies of butterflies, I next take up the relationship between fore- and hindwing patterns, population dynamics of the mimetic butterfly *Papilio polytes*.

8.3.1. Fore- and Hindwing Patterns

One of the most striking phenomena about wing color patterns is the close match between patterns of fore- and hindwing when wings are held at their resting position (Figure 8). This phenomenon is known as the Oudeman's principle (Oudeman, 1903). The overall unified pattern of both wings is usually explained by their functional significance, that is, the adaptation to the environment.

Figure 8. The close match between color patterns of fore- and hindwings (of *L. japonica*). White and black color bands running in the anterior-posterior direction of fore- and hindwings look like to be closely related to each other.

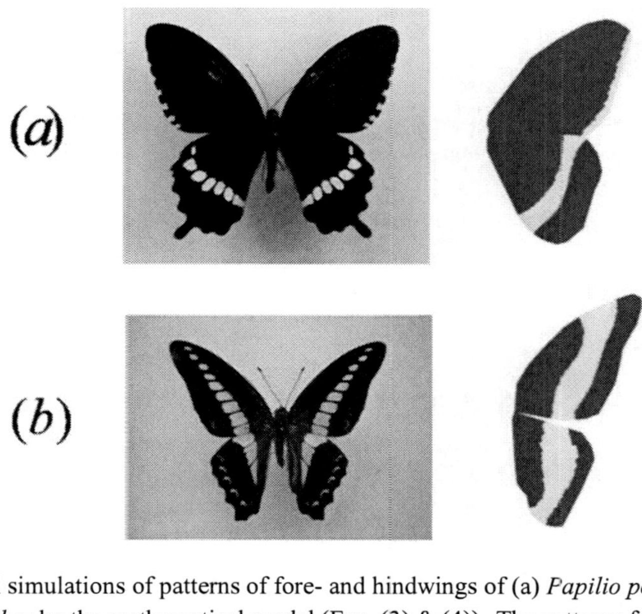

Figure 9. Numerical simulations of patterns of fore- and hindwings of (a) *Papilio polytes* & (b)*Graphium sarpedon* by the mathematical model (Eqs. (3) & (4)). The patterns formed on the forewing need not correlate to those of the hindwing in the sense that the formation mechanism is the same for both patterns. The independence of both patterns means that the coordination of unified patterns of fore- and hindwings is accidental, but not adaptive.

Sekimura et al. (2007) investigated the relationship between fore- and hindwings through mathematical modeling and computational analysis of *P. dardanus* and *P. polytes*. One of the main results is that the patterns formed on the forewing need not correlate to those of the hindwing in the sense that the formation mechanism is the same for both patterns (see Figure 9). The independence of both patterns means that the coordination of unified patterns of fore- and hindwings is accidental, but not adaptive. This is remarkable, because from Oudeman's principle, patterns appearing on the exposed surface of fore- and hindwing at the natural resting position are often integrated to form a composite and unified adaptive pattern with their surrounding environment. However, the result does not always mean that coordinated unified patterns of fore- and hindwings are not affected by their surrounding environment.

8.3.2. Population Dynamics of P. Polytes

P. polytes is a species of swallowtail butterfly widely distributed across the Oriental region including the Sakishima Islands (southern islands group of Japan). It is known that the female is polymorphic with four forms (f. *cyrus*, f. *polytes*, f. *romulus*, and f. *theseus*), whereas the male is always monomorphic. In the Sakishima Islands, however, there exist only two female forms, f. *cyrus* (non-mimic) and f. *polytes* (mimic). The form f. *cyrus* resembles the male, and the other form f. *polytes* resembles an unpalatable butterfly *Pachliopta aristolochiae* (model). Interestingly, in their distribution range, particular islands do not have *P. aristolochiae* and f. *polytes*, or *P. aristolochiae*.

Based on published records, Uesugi (1991) reviewed temporal changes in the occurrence of f. *polytes* and its model *P. aristolochiae* in the Sakishima Islands. He found that there had not been any records of *P. aristolochiae* in the Sakishima Islands from 1907 to 1967, but it was first recorded in 1968 in the Yaeyama Islands, the southernmost islands group of the Sakishima Islands and the records of f. *polytes* (mimic) increased rapidly after 1972 there. He also found that on Miyako-jima Island, a northern island of the Sakishima Islands, the number of the mimic butterflies has increased after establishment of the model butterfly in 1975. The increase in the population of f. *polytes* was discussed and explained in the context of adaptive significance of the female-limited Batesian mimicry (Uesugi, 1996).

Recently, Sekimura et al. (2013) have investigated the population dynamics of f. *polytes* by use of mathematical modeling, that includes four major variables, i.e., population densities of three kinds of butterfies (two female forms f. *cyrus*, f. *polytes* and the unpalatable butterfly *P. aristrochiae*) and their predator (submitted).

9. Integration of Key Aspects of Diversity in Wing Patterns

In order to understand how the diversity in wing patterns has been derived, we first have to analyze specific aspects (e.g., genetic aspect, ecological aspect, evolutionary aspect, and so on) underlying the pattern formation mechanisms and evolution. We then have to integrate them by new comprehensive methods including mathematical and computational analyses. A hint or way how to do this is for us to understand that the butterfly wing is an organ of a living butterfly surviving under various kinds of environmental restrictions such as various natural conditions, interspecific and intraspecific competition, and so on. The matter of utmost concern for a butterfly would be to survive in a given environment, and most of its characteristics including wing color pattern would have evolved to fit or adapt to the surrounding environment. This means that such characteristics including wing color pattern represent some kinds of strategies for survival, that must be ultimately related to changes in the fitness in the population biology. Here, the word "fitness" roughly means the number of offspring per adult in the next generation, a useful quantity to discuss temporal changes in population of an organism.

The problem is to find the way how to integrate key aspects. As an example, I would like to introduce a theoretical research by Ohsaki (1995), who succeeded to link the biochemical aspect to mimicry of a female-limited mimetic species *P. polytes*, whose females mimic the wing-color patterning of the unpalatable *P. aristrochiae* by means of the chemical pigment carotenoid It is known that the physiological life span of the mimic f. *polytes* is shorter than the non-mimic f. *cyrus,* In general, wing color patterns have specific biological functions, such as mutual recognition, protection from predators, and mimicry, which play important roles in butterfly survival. On the other hand, chemical pigments of wing colors are synthesized in cells by biochemical reaction networks. By use of the cost-benefit theory, Ohsaki linked the chemical pigment to mimicry, and he concluded that owing to their consumption of the chemical, the physiological life span of mimetic females is shorter than that of non-mimetic females. This conclusion may indicate a reduction in the fitness of the mimetic females. This example shows that changes in wing color patterning could affect the fitness or the future population of a butterfly species, and at the same time, this shows a deep linkage of a biochemical process to a different ecological event, mimicry.

Conclusion and Discussions

In this article, I briefly reviewed the progress of mathematical models for formation and diversification of color pattern in butterfly wings. I also reviewed another kind of mathematical model for cellular pattern formation of scale cells as the basic process of pattern formation in butterfly wings, even though both color pattern and cellular pattern are not directly related with each other. The mathematical models are all constructed on the basis of experimental data, which involve evidences in both macro- and micro-level scales, depending on pattern forming phenomena.

Until today, genetic or molecular level researches have focused mainly on the local pattern such as eyespot pattern, and on the other hand, there exist few researches on the global pattern. With respect to mathematical modeling, the situation is almost the same as that in experimental researches. It is generally recognized that in order to understand evolution and diversity in biological phenomena including the butterfly wing pattern, we have to investigate non-model organisms such as insects including butterfly from both experimental and mathematical modeling aspects.

There exist discussions on the general concept and methodology for the treatment of morphological phenomena and evolution in biology and paleontology such as "morphodynamics" (Seilacher, 1991). Regarding butterfly wing patterns, I have stressed the importance of integration of different aspects such as gene and environment for full understanding of the diversity in wing color patterns (Sekimura, 2013). A hint to go ahead is for us to simply remember that the butterfly wing is an organ of a living organism, a butterfly surviving in a variety of surrounded environments. Different aspects on the color pattern such as gene and environment are from characteristics of a single living organism, a butterfly, and they should be unified into one comprehensive feature of the butterfly after integration process. However, it is not always clear how to pursue the integration process at the moment.

I think that at least we should do the process of linking one aspect to another. I would like to show the reader an example, even though the animal in question is not butterfly, but fish. This is an experiment-based theoretical paper (Asai et al., 1999), in which the analysis has been done by using reaction-diffusion equations, that is, an activator-inhibitor system. Parameters such as reaction rates and diffusion constants in the equations originate from basic biochemical reaction networks and random mobilities of chemicals in cells. Authors succeeded to link parameter values in the activator-inhibitor system to activities of the *leopard* gene, which controls changes in Zebrafish skin color patterns. This is a good case study to connect a biochemical process directly to genetic activity.

We are at the first stage in the course of an integrative approach to the analysis of pattern and diversity in butterfly wings. We could connect a few of all the different aspects that are necessary to catch a full understanding of evolution and diversity in wing patterns. All of these step by step processes will be integrated into a comprehensive and realistic feature of wing patterns under the concept of survival strategies of a beautiful creature, butterfly. Some mathematical models have contributed to our understanding of the problem so far and, in the future, they will continue to play an important role in the analysis of pattern formation and the integration process of different aspects in question.

ACKNOWLEDGMENTS

First of all, I would like to thank the editor of this book for inviting me to write a review article on the progress of mathematical models for color pattern formation and diversity in butterfly wings. I also thank Prof. Fred Nijhout of Duke University for introducing me to the wonderful and beautiful world of butterflies. Finally, I thank all my collaborators for helping and encouraging me to continue working on this fantastic theme.

REFERENCES

Asai R., Taguchi, E., Kume,Y., Saito, M., & Kondo,S. (1999). Zebrafish leopard gene as a component of the putative reaction–diffusion system. *Mech. Dev.*, 89, 87–92.

Bard, J.B.L., & French, V. (1984). Butterfly wing patterns: how good a determining mechanism is the simple diffusion of a single morphogen?, *J. Embryol. exp. Morph.*. 84, 255-274.

Brakefield, P. B., Gates, J., Keys, D., Kesbeke, F., Wijingaarden, P. J., Monteiro, A., French, V., & Carroll, S. B. (1996). Development, plasticity and evolution of butterfly eyespot patterns, *Nature*, 384:236-242.

Brunetti, C.R., Selegue, J.E., Monteiro, A., French, V., Brakefield, P.B., & Carroll, S.B. (2001). The generation and diversification of butterfly eyespot color patterns. *Curr. Biol.* 11,1578–1585.

Carroll, S.B., Gates, J., Keys, D.N., Paddock, S.W., Panganiban, J.E., Selegue, J.E., & Williams, J.A. (1994). Pattern formation and eyespot determination in butterfly wings, *Science* 265,109-114.

Clark, R., Brown, S. M. Collins, S. C. Jiggins, C. D. Heckel, G. D. & Vogler. A. P. (2008).

Colour pattern specification in the Mocker swallowtail *Papilio dardanus*: the transcription factor *invected* is a candidate for the mimicry locus H, *Proc. Roy. Soc. London,* B, 275:1181-1188.

Clarke, C. A. & Sheppard, P. M. (1959). The genetics of some mimetic forms of *Papilio dardanus,* Brown, and *Papilio dardanus,* Linn, *J. Genet*. 56:237-259.

Clarke,C. A. & Sheppard, P. M. (1960). The evolution of mimicry in the butterfly *Papilio dardanus, Heredity*, 14:163-173.

Dilão, R. & Sainhas, J. (2004). Modelling butterfly wing eyespot patterns, *Proc. R. Soc. Lond.*, B, 271, 1565-1569.

Evans, T.M. & Marcus, J.M. (2006). A simulation study of the genetic regulatory hierarchy for butterfly eyespot focus determination. *Evol. Dev.* 8(3),273-283.

Geirer, A. & Meinhardt, H. (1972). A theory of biological pattern formation. *Kybernetik*, 12, 30-39.

Koch, P. B. & Nijhout, H. F. (2002), The role of wing veins in colour pattern in the butterfly *Papilio xuthus* (Lepidopteran: Papilionidae)*, Eur. J. Entomol*. 99:67-72.

Monteiro, A., Glaser, G., Stockslager, S., Glansdorp, N., & Ramos, D. (2006). Comparative insights into questions of lepidopteran wing pattern homology. *BMC Developmental Biology,* 6, 52.

Murray, J.D. (1981). On pattern formation mechanisms for lepidopteran wing patterns and mamalian coat markings, *Phil. Trans. R. Soc. Lond.*, B 295, 473-496.

Murray, J.D. (2003). Mathematical Biology II: Spatial models and biomedical application, (3rd edition), Springer-Verlag, Berlin Heidelberg.

Nardi, J.B. & Magee-Adams, S.M. (1986). Formation of scale spacing patterns in a moth wing. I. Epithelial feet may mediate cell rearrangement. *Dev. Biol.*, 116, 278-290.

Nardi, J.B. (1992). Dynamic expression of a cell surface protein during rearrangement of epithelial cells in the *Manduca* wing monolayer. *Dev. Biol.*, 152,161-171.

Nardi, J.B. (1994). Rearrangement of epithelial cell types in an insect wing monolayer is accompanied by differential expression of a cell surface protein. *Dev. Dynamics*, 199, 315-325.

Nijhout, H.F. (1980). Ontogeny of the color pattern on the wing of *Precis coenia* (Lepidoptera: Nymphalidae), *Dev. Biol.*, 80, 275-288.

Nijhout, H. F. (1981). The color patterns of butterflies and moths. *Scientific American,* 245, 145-151.

Nijhout, H. F. (1990). A comprehensive model for color pattern formation in butterflies. *Proc. R. Soc. Lond.*, B 239,81-113.

Nijhout, H. F. (1991). The Development and Evolution of Butterfly Wing Patterns. Smithsonian Institution Press, Washington and London.

Nijhout, H. F. (1994). Genes on the wing. *Science*, 265, 44-45.

Nijhout, H.F. (2003). Polymorphic mimicry in *Papilio dardanus*: mosaic dominance, big effects, and origins. *Evol. Dev.* 5, 579-592.

Nijhout, H.F., Maini, P.K., Madzvamuse, A., Wathen, A.J., & Sekimura, T. (2003). Pigmentation pattern formation in butterflies. *C.M. Biologies*, 326, 717-727.

Ohsaki, N. (1995). Preferential predation of female butterflies and the evolution of batesian mimicry. *Nature*, 373:173-175.

Oudeman, J.T. (1903). Etudes sur la position de repos chez les Lepidopteres. *Verhandelingen der Koning-klijke Akademie van Wetenschappen*, 10, 1-90.

Reed, R. D., Macmillan, W. O., & Nagy, L. M. (2008). Gene expression underlying adaptive variation in *Heliconius* wing patterns: non-modular regulation of overlapping *cinnabar* and *vermilion* prepatterns. *Proc. Roy. Soc. London,* B, 275:37−45.

Reed, R. D. & Gilbert, L. E. (2004). Wing venation and Distal-less expression in *Heliconius* butterfly wing pattern development. *Dev. Genes. Evol.* 214:628−634.

Schwanwitsch, B.N.(1924). On the ground plan of wing-pattern in nymphalids and certain other families of rhopalocerous Lepidoptera. *Proc. Zool. Soc. Lond.*, B 34, 509-528.

Seilacher, A. (1991). Self-organizing mechanisms in morphogenesis and evolution. In: Contructional Morphology and Evolution, (eds. Schmidt-Kitter, N. & Vogel, K.), pp.251-271, Springer-Verlag, Berlin Heidelberg.

Sekimura, T., Maini, P.K., Nardi, J.B., Zhu M., & Murray, J.D. (1998). Pattern formation in lepidopteran wings. *Comments Theor. Biol.*, 5, No.2-4, 69-87.

Sekimura, T., Zhu, M., Cook, J., Maini, P.K., & Murray, J.D. (1999). Pattern formation of scale cells in lepidoptera by differential origin-dependent cell adhesion. *Bull. Math. Biol.*, 61, 807-827.

Sekimura, T., Madzvamuse, A., Wathen, A.J., & Maini, P.K. (2000). A model for colour pattern formation in the butterfly wing of *Papilio dardanus*, *Proc. R. Soc. Lond.*, B 267,851-859.

Sekimura, T., Madzvamuse, A., & Maini, P.K. (2007). Pigmentation pattern formation in butterfly wings: global patterns on fore- and hindwing. In Mathematical Modeling of Biological Systems, Volume I (eds. Deutsch, A., Brusch, L., Byrne, H., de Vries, G., and Herzel, H.), Chap.12, pp.141-148, In Series: *Modeling and Simulation in Science, Engineering and Technology*, Birkhauser, Boston Basel Berlin.

Sekimura, T. (2013). An Integrative approach to the analysis of pattern formation in butterfly wings −experiments and models−. In Pattern Formation in Morphogenesis, (eds. Capasso, V., Gromov, M., Harel-Bellan, A., Morozova, N., & Pritchard, L.L.), *Springer Proceedings in Mathematics* 15, pp.121-136, Springer-Verlag Berlin Heidelberg.

Sekimura, T., Fujihashi, Y., Takeuchi, Y. (2013). A model for population dynamics of the mimetic butterfly *Papilio polytes* in Sakishima Islands, Japan. (submitted for publication)

Süffert, F.(1927). Zur vergleichende analyse der schmetterlingszeichnung. *Biologisches zentralblatt*, 47, 385-413.

Turing, A.M. (1952). The chemical basis of morphogenesis, *Phil. Trans. Roy. Soc. Lond.* 237, 37-72.

Uesugi, K. (1991). Temporal changes in records of the mimetic butterfly *Papilio polytes* with establishment of its model *Pachliopta aristolochiae*, *Jpn. J. Ent.*, 59 (1):183−198.

Uesugi, K. (1996). The adaptive significance of Batesian mimicry in the swallowtail butterfly, *Papilio polytes* (Insecta, Papilionidae):associative learning in a predator. *Ethology*, 102:762−775.

Vane-Wright, R.I., Raheem, D.C., Cieslak, A., & Vogler, A.P. (1999). Evolusion of the mimetic African swallowtail butterfly *Papilio dardanus*: molecular data confirm relationship with *P. phocus* and *P. constantius*. *Biol. J. Linn. Soc.*, 66, 215-229.

Yoshida, A. & Aoki, K. (1989). Scale arrangement pattern in a lepidopteran wing. I. Periodic cellular pattern in pupal wing of *Pieris rapae*. *Develop. Growth Differ.*, 31, 601-609.

In: Advances in Zoology Research. Volume 6
Editor: Owen P. Jenkins
ISBN: 978-1-61122-061-2

Chapter 7

D-Amino-Acid Oxidase in Rodents

Hiroko Ohide, Masahiro Yamanaka and Ryuichi Konno*
Department of Pharmaceutical Sciences, International University of Health and Welfare, Kitakanemaru, Ohtawara, Tochigi, Japan

Abstract

D-Amino-acid oxidase (DAO) catalyzes oxidative deamination of unnatural D-amino acids, producing the corresponding 2-oxo acids, hydrogen peroxide and ammonia. DAO is present in a wide range of organisms from yeasts to humans. In mammals, it is present in the brain, liver and kidney. The number of amino acid residues consisting of DAO is 347 in the human, pig, and rabbit. The guinea pig DAO has 347 amino acid residues whereas DAOs of the rat and hamster consist of 346 amino acid residues. Furthermore, mouse DAO has 345 amino acid residues. Thus, the numbers of amino acid residues are different depending on the species of rodents. Accordingly, we searched mammalian DAOs, especially rodent DAOs, in databases and examined their amino acid sequences. We found that DAOs of monkey, cattle, squirrel, and degu have 347 amino acid residues and jerboa DAO has 346 residues whereas the prairie vole DAO has 345 residues. Comparison of amino acid sequences has indicated that the number of the amino acid residues and the homology of amino acid sequences are clearly parallel to the taxonomic positions of the rodents.

Introduction

D-Amino-acid oxidase (DAO, EC 1.4.3.3) catalyzes oxidative deamination of D-amino acids, stereoisomers of naturally occurring L-amino acids. D-Amino acids are converted to the corresponding 2-oxo acids, hydrogen peroxide and ammonia. DAO has wide substrate specificity, so many neutral and basic D-amino acids are catabolized (Dixon and Kleppe 1965). This enzyme was found in 1935 (Krebs 1935). DAO is present in a wide range of organisms from yeasts to mammals. In mammals, it is present in the kidney, liver and brain

* Corresponding author: Ryuichi Konno (konno-r@iuhw.ac.jp, FAX: +81-287-24-3115).

(see Ohide et al. 2011). The physiological role of DAO is the metabolism of external and internal origins of D-amino acids. Therefore, large amounts of D-amino acids accumulate in the organs and body fluid of the mutant mice and rats, which lack DAO (see Konno et al. 2010; see Yamanaka et al. 2012). In the brains of mammals, D-serine is abundantly present and functions as a co-agonist of NMDA receptors, a subtype of glutamate receptors. Since DAO reduces this D-serine level, it works as a modulator of neurotransmission. Therefore, disorders in DAO cause neuropsychiatric diseases such as schizophrenia (see Verrall et al. 2010) and amyotrophic lateral sclerosis (Sasabe et al. 2012).

DAO is an enzyme that consists of a single polypeptide and a coenzyme, flavin adenine dinucleotide (FAD). The number of the amino acid residues that make up a DAO polypeptide depends on the animals. Human DAO consists of 347 amino acid residues with a molecular mass of 39,474 Da (Momoi et al. 1988). Pig and rabbit DAOs (Fukui et al. 1987; Momoi et al. 1990) also have 347 amino acid residues. On the contrary, the number of amino acid residues is different depending on the species of the rodents. The guinea pig DAO has 347 amino acid residues (Konno et al. 1999). Rat and hamster DAOs (Konno 1998; Tsuchiya et al. 2003) have 346 amino acid residues. Furthermore, the mouse DAO has 345 amino acid residues (Tada et al. 1990). These differences prompted us to search the database for mammalian DAOs, especially rodent DAOs. We examined the number of amino acid residues of their DAOs and compared their amino acid sequences in detail.

PROCEDURES

We searched mammalian DAOs for the compositions and amino acid sequences that were deposited in databases (National Center for Biotechnology Information). The amino acid sequence of DAO of the gerbil (*Meriones unguiculatus*) was determined as described before (Konno 1998; Konno et al. 1999; Tsuchiya et al. 2003). Amino acid sequences of various animals were aligned and compared by pair. The number of identical amino acid residues was counted and this number was divided by the number of total amino acids. The total amino acids were 347 in most cases but 346 in the case of the gerbil, hamster, and rat. The total number was 345 in the case of the prairie vole and mouse. The quotient was multiplied by 100 to show the percentage.

RESULTS AND DISCUSSION

We searched for amino acid sequences of DAOs of various mammals that were deposited in the database. It was found that DAOs of monkey (*Macaca fascicularis*), cattle (*Bos taurus*), squirrel (thirteen-lined ground squirrel, *Spermophilus tridecemlineatus*), and degu (*Octodon degus*) have 347 amino acid residues as do DAOs of the human, pig, and rabbit (Table 1).

Table 1. Animals used in the analysis of amino acid sequence of DAO

Common name	Scientific name	Accession number	Amino acid residues
Human	*Homo sapiens*	NM_001917	347
Monkey	*Macaca fascicularis*	NM_001283949	347
Cattle	*Bos taurus*	NM_001206225	347
Pig	*Sus scrofa*	NM_214066	347
Rabbit	*Oryctolagus cuniculus*	NM_001082189	347
Squirrel	*Spermophilus tridecemlineatus*	XM_005340103	347
Guinea pig	*Cavia porcellus*	AJ007710	347
Degu	*Octodon degus*	XM_004636274	347
Jerboa	*Jaculus jaculus*	XM_004664104	346
Hamster	*Mesocricetus auratus*	AB016530	346
Prairie vole	*Microtus ochrogaster*	XM_005344251	345
Rat	*Rattus norvegicus*	NM_053626	346
Mouse	*Mus musculus*	NM_010018.2	345

DAO of the jerboa (lesser Egyptian jerboa, *Jaculus jaculus*) was found to consist of 346 amino acids like the hamster and rat DAOs. It was surprising that the DAO of the prairie vole (*Microtus ochrogaster*) had 345 amino acid residues, because only mouse DAO was known to possess 345 amino acids so far. The Mongolian gerbil DAO was determined to have 346 amino acid residues. DAOs of human and rat were compared. Amino acid sequences were aligned (Figure 1). Human DAO had 347 amino acid residues whereas rat DAO had 346. The 27th amino acid residue was missing in rat DAO (Konno 1998; Tsuchiya et al. 2003). Overall, the amino acid sequences were very similar. As shown, 279 of 346 amino acids (80.6%) were identical (Figure 1 and Table 2).

```
a  MRVVVIGAGV IGLSTALCIH ERYHSVLQPL DIKVYADRFT PLTTTDVAAG LWQPYLSDPN   60
b  •••A•••••• •••••••••• ••••PA-••• HM•I•••••• •F••S••••• •••••••••S

a  NPQEADWSQQ TFDYLLSHVH SPNAENLGLF LISGYNLFHE AIPDPSWKDT VLGFRKLTPR  120
b  •••••E•N•• •••H•Q•CL• •••••KM••A ••••••••RD EV•••F••S• •••••••••S

a  ELDMFPDYGY GWFHTSLILE GKNYLQWLTE RLTERGVKFF QRKVESFEEV AREGADVIVN  180
b  ••••••••S• •••N•••L•• ••S••S•••• •••••••••I H•••A••••• V•G•V•••I•

a  CTGVWAGALQ RDPLLQPGRG QIMKVDAPWM KHFILTHDPE RGIYNSPYII PGTQTVTLGG  240
b  •••••••••• A•AS•••••• ••IQ•E•••I •••••••••S L••••••••• ••SK••••••

a  IFQLGNWSEL NNIQDHNTIW EGCCRLEPTL KNARIIGERT GFRPVRPQIR LEREQLRTGP  300
b  V••••••••• •SVH•••••• KS••Q••••• •••••M••L• ••••••••V• ••••R••F•S

a  SNTEVIHNYG HGGYGLTIHW GCALEAAKLF GRILEEKKLS RMPPSHL  347
b  •SA••••••• •••••••••• •••M•••N•• •K•••••N•• •••••••
```

Figure 1. Amino acid sequence of human and rat DAOs. Amino acid is shown in one-letter. Amino acid sequences were aligned. The same amino acid residue in rat DAO is shown in dot. Human DAO had 347 amino acid residues whereas rat DAO had 346. The 27th amino acid residue was missing in rat DAO. a: human, b: rat.

Table 2. Homology of amino acid sequence of DAO between two animals

	Human	Monkey	Cattle	Pig	Rabbit	Squirrel	Guinea pig	Degu	Jerboa	Hamster	Prairie vole	Gerbil	Rat
Monkey	93.9												
Cattle	83.3	80.7											
Pig	84.7	83.9	89.9										
Rabbit	83.6	83.3	80.1	80.4									
Squirrel	81.0	78.7	77.2	78.7	77.5								
Guinea pig	85.6	84.4	81.6	84.1	81.8	79.3							
Degu	84.1	84.1	80.1	81.6	80.7	79.3	90.5						
Jerboa	80.6	80.1	77.5	77.7	78.6	75.7	81.5	79.5					
Hamster	80.3	79.5	74.6	77.5	76.9	74.6	80.6	78.9	78.6				
Prairie vole	80.3	79.4	75.1	78.0	77.7	74.2	79.7	78.6	78.6	90.0			
Gerbil	82.1	80.9	76.9	78.6	78.9	76.3	80.9	80.3	82.4	88.7	87.8		
Rat	80.6	79.5	76.9	78.0	79.2	76.3	80.1	80.6	80.3	87.6	86.4	89.0	
Mouse	80.9	79.1	75.4	78.6	78.8	76.2	80.9	80.6	82.3	89.3	87.0	89.9	93.6

The number indicates the percentage of identical amino acid residues consisting of DAOs of two animals.

```
a  MRVVVIGAGV IGLSTALCIH QRYHSDLQPL DMKVYADRFT PHTTTDVAAG LIQPYISPPE  60
b  •••••••••• •••••••••C E••••V••Q• •LR••••••• •L•N•••••• •W•••L•G•D
c  •••••••••• •••••••••Y E••••I••S• •L•••••••• •L•N•••••• •W•••L•G•S
d  •••••••••• •••••••••• E••R•V-••• SVQ••••H•• •L••S••••• •C•••L•T•S
e  •••••••••• •••••••••• E•FSPV-••• H••I•••••• •F••S••••• FW•••L•D•R
f  •••••••••• •••••••••• E•F•PA-••• HL•••••••• •F••S••••• •W•••L•DSS
g  •••••••••• •••••••••• E•F•PT-L•• HV•••••••• •L••S••••• FW•••L•D•I
h  •••A•••••• •••••••••• E•••PA-••• H••I•••••• •F••S••••• •W•••L•D•S
i  •••A•••••• •••••••••• E•••PT-••• H••I•••••• •F••S••••• •W•••L•D•S

a  DPREMDWNQQ TFDYLLSQVH SPTADKMGLA LISGYNLFRE AVPDPVWKDI VLGFRKLTPR 120
b  N•Q•VE•••• •••••••HI• ••N•EQ•••• •V•••••••• •••••F••NM ••••••••••
c  N•Q••A•••• •••••••HIY ••N•E••••• ••••••••H• •••••F••N• ••••••••••
d  N•Q•AE•S•• •••••••HL• ••N••••••• •••••••••D T•QE•F•••T •••••E••••
e  N•Q•VE•••• •••••••HI• ••N•E••••S ••••••••K• E••••F•RNT ••••••••••
f  N•Q•AE•G•• •••••••HI• •QD•E••••S •V••••••QH E••••S•RNT •••••R••••
g  N•Q•A••••• •••••••HL• ••D•E••••• •••••••••D E••••F••NT ••••••••••
h  N•Q•AE•••• •••H•Q•CL• ••N•E••••• •••••••••D E••••F••ST •••••••••S
i  N•Q•AE•S•• •••••••CL• ••N•E••••• •••••••••D E••••F••NA •••••••••S

a  ELDMFPDYGY GWFNTSLILD GRTYLKWLTE RLTERGVQFF LRKVASLEEV ARGGVDVIIN 180
b  •••V•••••• •••H••••IE •KS••A•••• •••••••K•• Q•••E••••• ••••A•••••
c  •••V••••S• •••H••••IE •KS••E•••• •••••••K•• Q•••E••••• ••••••••••
d  •••I••••S• •••H•••••E •KA••Q•••• •••••••K•I Q•••E••••• ••••A•••••
e  •M•I•••••• •••••••T•E •KS••P•••• •••••••KL• H•••E•F••• ••••A•••••
f  •M•I••E•S• •••••••S•E •KR••P•••• •••••••KL• H•••E•F••• ••••A•••••
g  •M•I••••S• •••••••L•E •KH••P•••• ••R••••KLI H•••E•F••• ••••A•••••
h  ••••••••S• •••••••L•E •KS••S•••• •••••••K•I H•••••F••• V•••••••••
i  •M•L•••••• •••••••L•E •KS••P•••• •••••••KLI H•••E••••• ••-•••••••

a  CTGVWAGALQ PDPLLQPGRG QVIQVEAPWM KHFIITHDPE RGIYKTPYII PGSQTVTLGG 240
b  •••••••••• •••••••••• •I•K•N•••I ••••L••••• •••••S•••• ••I•E•••••
c  •••••••••• •••••••••• •I•K•N•••I ••••V••NLK S•••TS•••• ••M•••••••
d  •••••••••K ••AQ•••••• •••R•D•••• •••••••••A Q•••TS•••• ••IHE•••••
e  •••••••••• A•TS•••••• •I•••••••• ••••L••••R L•••NS•••• •••K••••••
f  •••••••••• T•T••••••• •I•••••••• ••••L•••SS ••••NS•••• •••K••••••
g  •••••••••• A••S•••••• •I•••D•••• •••••••••V H•••NS•••• •••K••••••
h  •••••••••• A•AS•••••• •I•••••••I ••••L••••S L•••NS•••• •••K••••••
i  •••••••••• A•AS•••••• •I•••••••I ••••L••••S L•••NS•••• •••K••••••

a  TFQLGNWTEL IDTQDHDTIW EGCCRLEPTL KNARLVGVRA GFRPARPRIR LERDRLPGGP 300
b  I••••••N•I NS••••N••• K•••S••••• R•••I••EYT ••••V••QL• •••EQ•RV•S
c  IN•••••S•I NS••••N••• •••••••••• ••••IA•EHT ••••V••QL• ••••Q•RV•S
d  I••••••S•V KSV•••N••• K•••K•A••• ••••I••EYV ••••V••QV• •••EW•HA•S
e  V••••••N•• NSVH••N••• KS••K••••• •••KI••ELT ••••V•HQV• •KKKQ•HF•S
f  V••••••N•• NSVH••NS•• K•••Q••••• ••••I••ELT ••••V••QV• •••EK•HF•S
g  V••••••S•• NSV•••N••• KA••K••••M ••••I••ELT ••••V••QV• •••E••RF•S
h  V••••••S•• NSVH••N••• KS••Q••••• ••••IM•ELT ••••V••QV• •••E••RF•S
i  I••••••SG• NSVR••N••• KS••K••••• ••••I••ELT ••••V••QV• •••EW•HF•S

a  ADTEVIHNYG HGGQGITIHW GCALEAAKLF GKILEERKLS RTPPSHL  347
b  •N•••••••• •••Y•L•••• •••••••••• ••••••KE•• •M•••••
c  •S•••••••• •••Y•L•••• •••••••N•• ••••••KE•Y •M•••••
d  LKA••••••• •••Y•L•••• •••••••R•• ••••••K••• SI••P••
e  SSV••••••• •••Y•L•••• •••M•••N•• ••••••K••• •M•A•••
f  SSA••••••• •••Y•V•V•• •••M•••N•• ••••••K•-• •M•••••
g  SSA••••••• •••Y•L•••• •••M•••N•• ••••••K••• •MT••••
h  SSA••••••• •••Y•L•••• •••M•••N•• ••••••KN•• •M•••••
i  SSA••••••• •••Y•L•••• •••M•••N•• ••••••K••• •L•••••
```

Figure 2. Amino acid sequences of DAOs of eight rodents. The same amino acids of DAOs other than squirrel were shown in dot. Missing amino acid residue is shown by a hyphen. a: squirrel, b: guinea pig, c: degu, d: jerboa, e: hamster, f: prairie vole, g: gerbil, h: rat, i: mouse.

Subsequently, we compared amino acid sequences of various DAOs pair-wise. Table 2 shows the percentage of identical amino acid residues in DAOs of two animals. The primary structures of mammalian DAOs were homologous in general and more than 70% of amino acid residues were identical. Among these, highly homologous pairs were observed. 93.9% of amino acid residues of human and monkey DAOs were identical. It is reasonable because these animals belong to the same taxonomic Order, Primates. Similarly, cattle and pig DAOs were highly homologous (89.9%). These animals are in the same Order, Cetartiodactyla. However, homology in the amino acid sequences of DAOs of the animals belonging to the

Order Rodentia was not as high as those in other Orders. This may be because a wide variety of rodent species were generated and have prosperously spread all over the world as the result of adaptive radiation.

Then, amino acid sequences of DAOs of eight species of Rodents were examined in detail (Figure 2). The primary structures of rodent DAOs are basically similar. Highly conserved regions spread throughout the molecule. Amino acid 3-17, 37-38, 44-45, 312-317 are known to be the binding regions for coenzyme, FAD. The sequence, GXGXXG, at the N terminus is the consensus sequence for FAD binding (Pilone 2000). Amino acid 164 (V) and 182 (T) serve as the FAD binding sites. Amino acid 53 (Q), 217 (H), 228 (Y), 283 (R), and 313 (G) are determined as the substrate-binding sites.

Homology of amino acid sequence of DAOs was found to be clearly parallel to the taxonomic positions of rodents (Table 2 and Figure 3) (Blanga-Kanfi et al. 2009; Churakov et al. 2010). Taxonomically, the Order Rodentia is divided into three Suborders, Sciuromorpha (Sciurognathi), Hystricomorpha (Ctenohystrica), and Myomorpha. Squirrel belongs to Suborder, Sciuromorpha. Degu and guinea pig belong to the second Suborder, Hystricomorpha. Jerboa, hamster, prairie vole, rat, and mouse belong to the third Suborder, Myomorpha.

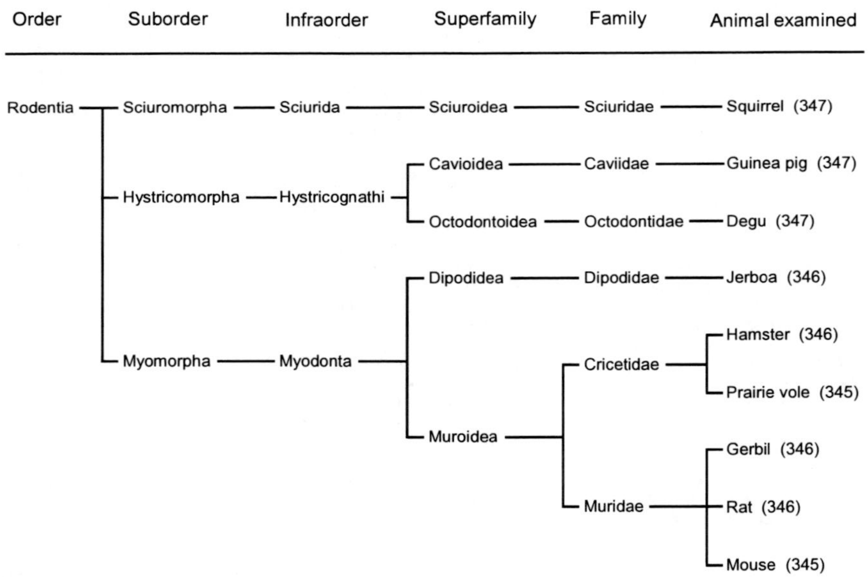

Figure 3. Taxonomic relationship of the rodent animals used for the analysis of amino acid sequence of DAOs. The figure in the parentheses indicates the number of amino acid residues consisting of DAO.

Homology of amino acid sequences of DAOs was not high when the animals belonging to three different Suborders were compared (Table 2). DAOs of squirrel, guinea pig, and degu had 347 amino acid residues whereas DAOs of rodents in the Suborder Myomorpha commonly lacked the 27th amino acid residues (Figures 2 and 3).

The amino acid sequence of squirrel DAO is different from that of other rodent DAOs (70% level of identity) (Table 2). Amino acid sequences of guinea pig and degu are very similar to each other (90.5%) but different from those of other rodents. Guinea pig and degu are separated at the level of Superfamily. Primary structure of DAO of jerboa was slightly different from that of DAOs of hamster, prairie vole, gerbil, rat, and mouse (Table 2). Jerboa is separated from the latter group at the level of Superfamily. Amino acid sequences of DAOs of hamster, prairie vole, gerbil, rat, and mouse, which belong to the same Superfamily, had high resemblance. Particularly, homology of rat and mouse DAOs was very high (93.6%).

These results indicate that the taxonomic relationship is strongly reflected on the primary structure of DAO. Therefore, composition and amino acid sequence of DAO, on the contrary, may be useful for the classification of rodents.

Rodents in Suborder Myomorpha commonly lacked the 27th amino acid residue (Figure 2). Therefore, this deletion seems to have occurred relatively early in ancestral animals after divergence of Myomorpha from Sciuromorpha and Hystricomorpha (Figure 3).

Jerboa DAO also lacked the 27th amino acid residue but the amino acid sequence was not as similar as that of DAOs of hamster, gerbil, rat, or mouse. One curious feature found in jerboa DAO is a PHL sequence at the carboxy terminus (Figure 2), whereas other mammalian DAOs have the SHL sequence at the end (Figure 2). The SHL sequence acts as the peroxisomal targeting signal (Gould et al. 1988). Therefore, it is interesting to examine whether jerboa DAO is present in peroxisomes of cells.

Mouse and prairie vole DAOs had 345 amino acid residues. They both lacked the 27th amino acid residue, as did other rodents in Myomorpha. In addition, the mouse DAO lacked the 173th amino acid residue whereas the prairie vole DAO lacked the 339th amino acid residue. Since they belong to different Families, these deletions seem to have occurred independently. Mechanisms underlying these deletions as well as the 27th deletion need to be clarified.

The physiological role of DAO in humans has not yet been fully understood. It is not clear how the disorders in DAO cause psychiatric diseases such as schizophrenia and amyotrophic lateral sclerosis in humans. A variety of DAOs in rodents may be helpful in solving these problems.

REFERENCES

Blanga-Kanfi S, Miranda H, Penn O, Pupko T, DeBry RW & Huchon D (2009) Rodent phylogeny revised: analysis of six nuclear genes from all major rodent clades. *BMC Evol Biol* 9:71.

Churakov G, Sadasivuni MK, Rosenbloom KR, Huchon D, Brosius J & Schmitz J (2010) Rodent evolution: back to the root. *Mol. Biol. Evol* 27:1315-1326.

Dixon M & Kleppe K (1965) D-Amino acid oxidase. II. Specificity, competitive inhibition and reaction sequence. *Biochim. Biophys Acta* 96:368-382.

Fukui K, Watanabe F, Shibata T & Miyake Y (1987) Molecular cloning and sequence analysis of cDNAs encoding porcine kidney D-amino acid oxidase. *Biochemistry* 26:3612-3618.

Gould SJ, Keller GA & Subramani S (1988) Identification of peroxisomal targeting signals located at the carboxy terminus of four peroxisomal proteins. *J. Cell Biol* 107:897-905.

Konno R (1998) Rat D-amino-acid oxidase cDNA: rat D-amino-acid oxidase as an intermediate form between mouse and other mammalian D-amino-acid oxidases. *Biochim Biophys Acta* 1395:165-170.

Konno R, Hamase K, Maruyama R & Zaitsu K (2010) Mutant mice and rats lacking D-amino acid oxidase. *Chem Biodivers* 7:1450-1458.

Konno R, Kurabayashi A, Tsuchiya M & Niwa A (1999) Guinea pig D-amino-acid oxidase cDNA and phylogenetic position. *DNA Seq* 10:85-91.

Krebs HA (1935) Metabolism of amino-acids: Deamination of amino-acids. *Biochem. J.* 29:1620-1644.

Momoi K, Fukui K, Tada M & Miyake Y (1990) Gene expression of D-amino acid oxidase in rabbit kidney. *J. Biochem.* 108:406-413.

Momoi K, Fukui K, Watanabe F & Miyake Y (1988) Molecular cloning and sequence analysis of cDNA encoding human kidney D-amino acid oxidase. *FEBS Lett* 238:180-184.

Ohide H, Miyoshi Y, Maruyama R, Hamase K & Konno R (2011) D-Amino acid metabolism in mammals: biosynthesis, degradation and analytical aspects of the metabolic study. *J. Chromatogr B* 879:3162-3168.

Pilone MS (2000) D-Amino acid oxidase: new findings. *Cell Mol Life Sci* 57:1732-1747.

Sasabe J, Miyoshi Y, Suzuki M, Mita M, Konno R, Matsuoka M, Hamase K & Aiso S (2012) D-Amino acid oxidase controls motoneuron degeneration through D-serine. *Proc. Natl. Acad Sci USA* 109:627-632.

Tada M, Fukui K, Momoi K & Miyake Y (1990) Cloning and expression of a cDNA encoding mouse kidney D-amino acid oxidase. *Gene* 90:293-297.

Tsuchiya M, Kurabayashi A & Konno R (2003) Hamster D-amino-acid oxidase cDNA: rodents lack the 27th amino acid residue in D-amino-acid oxidase. *Amino Acids* 24:223-226.

Verrall L, Burnet PW, Betts JF & Harrison PJ (2010) The neurobiology of D-amino acid oxidase and its involvement in schizophrenia. *Mol. Psychiatry* 15:122-137.

Yamanaka M, Miyoshi Y, Ohide H, Hamase K & Konno R (2012) D-Amino acids in the brain and mutant rodents lacking D-amino-acid oxidase activity. *Amino Acids* 43:1811-1821.

In: Advances in Zoology Research. Volume 6
Editor: Owen P. Jenkins
ISBN: 978-1-61122-061-2

Chapter 8

THE RACCOON DOG (*NYCTEREUTES PROCYONOIDES*) IN THE COMMUNITY OF MEDIUM-SIZED CARNIVORES IN EUROPE: ITS ADAPTATIONS, IMPACT ON NATIVE FAUNA AND MANAGEMENT OF THE POPULATION

***Kaarina Kauhala*[1,*] *and Rafał Kowalczyk*[2]**
[1]Finnish Game and Fisheries Research Institute, Turku, Finland,
[2]Mammal Research Institute, Białowieża, Poland

ABSTRACT

Raccoon dog (*Nyctereutes procyonoides*) is an alien carnivore in Europe. It is today widespread and common in many parts of the continent. Here we review its adaptations to various environmental and climatic conditions, impact on the native fauna and management of the populations. The raccoon dog is very adaptable due to its versatile diet, ability to gather large fat reserves in autumn and to hibernate in areas where winters are harsh. It also has higher than expected reproductive potential, which compensates the high mortality rate of juveniles. Raccoon dogs are monogamous and the male participates in pup rearing. Since no helpers have been observed, juveniles usually disperse during their first autumn and some of them wander far. In a colonizing population also adults have a tendency to wander long distances. Due to massive and dispersed introductions, raccoon dogs are characterized by high genetic variation.

Raccoon dogs fitted well into the guild of medium-sized carnivores and share their home ranges with badgers (*Meles meles*) and foxes (*Vulpes vulpes*), and especially raccoon dogs and badgers partly use the same habitats, such as deciduous forests and fields. Raccoon dogs often settle in active badger setts to avoid predation and unfavourable weather conditions. Raccoon dogs may partly overlap their diet with that of native carnivores. Raccoon dogs can spread diseases and parasites, such as rabies and sarcoptic mange, to other carnivores due to their frequent opportunities of contact.

* Email: kaarina.kauhala@rktl.fi.

The raccoon dog is a true omnivore and eats anything it can find and catch. Although it sometimes preys on birds, there is no firm evidence that it affects bird populations. The occurrence of birds in its diet shows great spatial and seasonal variation but in most areas is fairly low. Raccoon dogs may have, however, an impact on amphibian populations, especially on islands.

Raccoon dogs are actively hunted in many countries. In Finland, the annual hunting bag is around 160,000 individuals, i.e., about 50% of the autumn population. In spite of many campaigns to reduce the population, the annual hunting bag has steadily increased during the past two decades in Finland. The hunting bag has also increased exponentially in Germany since the early 1990s. Due to their adaptability, large distribution and secretive life, raccoon dogs are very difficult to eradicate after they have established in an area.

INTRODUCTION

The raccoon dog (*Nyctereutes procyonoides*), which originates in eastern Asia, was introduced by Russians mainly to the European parts of the former Soviet Union during the first half of the 20th century (Lavrov 1971, Helle & Kauhala 1991). Due to its adaptability, tendency to wander and high reproductive potential it expanded its range rapidly to many eastern and northern European countries. Today it is among the most numerous carnivores in many countries and it is still expanding its range westwards and southwards in Europe.

Being an alien carnivore, the raccoon dog has been accused to cause damage especially to native game bird populations. It has also been suspected to compete with native carnivores for the best habitats or food resources. The raccoon dog has also been recognized as an important vector of diseases and parasites in its new range. Especially hunters' organizations have claimed intensive control or even eradication of raccoon dog populations but, despite of several campaigns to reduce its numbers, its populations seem to be increasing in Europe.

In this chapter we review the raccoon dog's adaptability, i.e., the reasons behind its success in Europe, its impact on native fauna, its role as a vector of diseases and parasites and the possibility to control its populations.

1. AN ADAPTABLE CARNIVORE

1.1. Omnivory and Diet Variation

Although the raccoon dog is a member of Carnivora, its diet is highly versatile and it is a true omnivore. This can even be seen in its morphology (Bannikov 1964): it has a long gut and small skull and tiny canine and carnassial teeth compared e.g., to those of the red fox (*Vulpes vulpes*), another canid of similar size. Its molars, adapted to grind plant and insect material, are well developed (Kauhala et al. 1998a). Raccoon dog's teeth thus reflect its omnivorous diet, which varies according to the availability of different food items and thus shows spatial and seasonal variation (e.g., Nasimovič & Isakov 1985, Sutor et al. 2010). Accordingly, the food niche of the raccoon dog is much wider than that of other medium-sized carnivores (Jędrzejewska & Jędrzejewski 1998, Kauhala et al. 1998b).

In many areas of Europe small mammals constitute a major part of the raccoon dog's diet (Kolosov et al. 1965, Kauhala et al. 1993a, 1998b, Sutor et al. 2010; Figure 1). Besides small rodents, raccoon dogs commonly prey on shrews (*Sorex* spp): 13-23% of raccoon dog scats contained remains of shrews in summer in southern Finland (Kauhala & Auniola 2001). Larger mammals, such as hares (*Lepus* spp.) occur occasionally in the diet but were probably consumed as carrion.

The occurrence of birds in the raccoon dog's diet varies according to season and area and increases with latitude (Sutor et al. 2010). For instance, in an island in SW Finland, birds occurred in 61% of raccoon dog scats (Ihalainen 2012), whereas in Germany remains of wild birds and eggs were found only in 7.5% of raccoon dogs' stomachs in summer (Sutor et al. 2010). Birds occur more often in the diet in years when voles are scarce than during years when voles are abundant (Ivanova 1962, Judin 1977, Kobylińska 1996). Many of the larger birds (e.g., ducks) were probably eaten as carrion (Barbu 1972, Kauhala & Auniola 2001).

Carrion occurs in the diet especially in winter in areas (e.g., Belarus, Poland and Germany) where raccoon dogs do not hibernate (Jędrzejewska & Jędrzejewski 1998, Sidorovich et al. 2000, Sutor et al. 2010). Ungulate carcasses are an important food source for raccoon dogs in winter and may affect their survival during the harshest period of the year. Carrion and other waste (e.g., human refuse from compost heaps) occurred also in about 50% of stomachs in all seasons in Finland (Kauhala et al. 1993a).

Amphibians were common in the diet of raccoon dogs especially in wetland areas where the share of small mammals was smaller (Barbu 1972, Sutor et al. 2010). Even frogspawn was consumed by raccoon dogs (Ihalainen 2012). Frogs and reptiles were more common in the diet on Finnish mainland, where many vipers (*Vipera berus*) are killed by cars, than on islands where no roads exist (e.g., Kauhala et al. 1998b).

Fish occurred in the diet especially in Finland in late winter when raccoon dogs pick small fish from the ice where fishermen discarded them (ice-fishing is popular in Finland; Kauhala et al. 1993a). Fish were also eaten by raccoon dogs in a wetland area in Germany where 11% of stomachs contained remains of fish (Sutor et al. 2010).

Insects and other invertebrates are commonly found in raccoon dogs' diet in all areas (Sutor et al. 2010). In summer, almost every scat contains remains of invertebrates, especially Coleoptera and Hymenoptera (Kauhala et al. 1993a). The most common beetles in the diet were Carabidae, Geotrupidae and Silphidae (Ihalainen 2012). Due to the small size of invertebrates the frequency of occurrence may, however, exaggerate their importance for raccoon dogs.

Raccoon dogs consumed berries and fruits frequently all over Europe (Sutor et al. 2010). Fruits and berries, such as bilberries (*Vaccinium myrtillus*) and lingonberries (*V. vitis-idaea*), are important food sources in late summer and autumn when raccoon dogs are accumulating fat reserves for winter (Nasimovič & Isakov 1985, Kauhala et al. 1993a). Raccoon dogs frequently visit gardens where they consume fruits, such as *Prunus* spp., *Malus* spp., *Pyrus* spp. and *Vitis* spp. (Nasimovič & Isakov 1985, Kauhala 2009, Ihalainen 2012).

The raccoon dog's versatile diet is one important factor behind its success in Europe: raccoon dogs can always find something to eat despite of habitat and season. The exception is harsh winters with thick and soft snow cover. In these conditions raccoon dogs are not able to move around and find food. They have solved the problem by gathering large fat reserves in autumn and hibernating in areas with harsh winters.

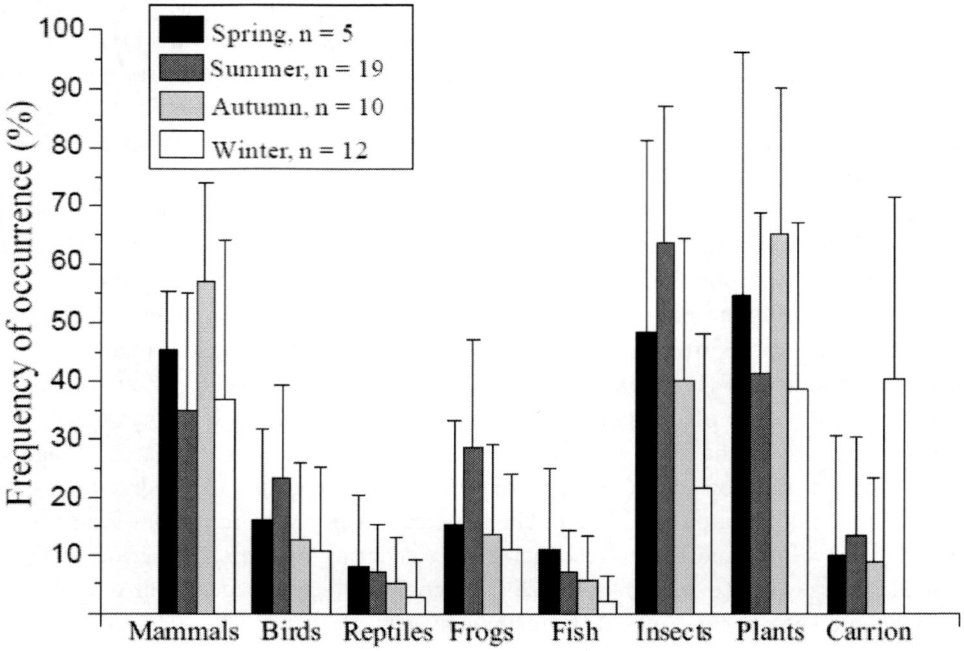

Figure 1. The diet of raccoon dogs in Europe in different seasons (pooled data from several areas). N indicates the number of data sets. Frequency of occurrence gives the percentage stomachs/feces which contained each food item. 'Insects' may contain some other invertebrates. Original figure in Kauhala 2009.

1.2. Hibernation and Fat Reserves

The raccoon dog is the only canid that hibernates in winter in northern areas. In southern Finland, hibernation usually starts in November and ends in March, i.e., during the season when food resources are sparse and difficult to find (Kauhala et al. 2007). Most raccoon dogs sleep when the air temperature is below -10 °C, the snow depth is > 35 cm and day length is < 7 hours, and most are active when the ambient temperature is above zero, there is no snow and the day length is > 10 hours (Kauhala et al. 2007). Raccoon dogs often wake up temporarily during winter and change the dens a few times. Raccoon dogs do not hibernate in more southern areas, e.g., in Germany (Drygala et al. 2008a). When hibernating, they settle in good shelters such as badger setts or hollow trees (Kauhala et al. 2007, Kowalczyk et al. 2008, Kowalczyk and Zalewski 2011).

Raccoon dogs accumulate large fat reserves and almost double their weight between early summer and late autumn (Kauhala 1993). Adults fatten themselves earlier than juveniles and reach their maximum mass and fat reserves in October-November, whereas juveniles reach their maximum mass one or two months later (Figure 2). Juveniles reach the adult body length in late autumn, usually in November (Kauhala 1993). Raccoon dogs lose about 43% of their body mass during hibernation (Mustonen et al. 2007). Fat accumulation in autumn and hibernation are regulated by hormonal chances, especially by thyroid hormone level, which is low during hibernation (Korhonen 1987, Nieminen et al. 2002).

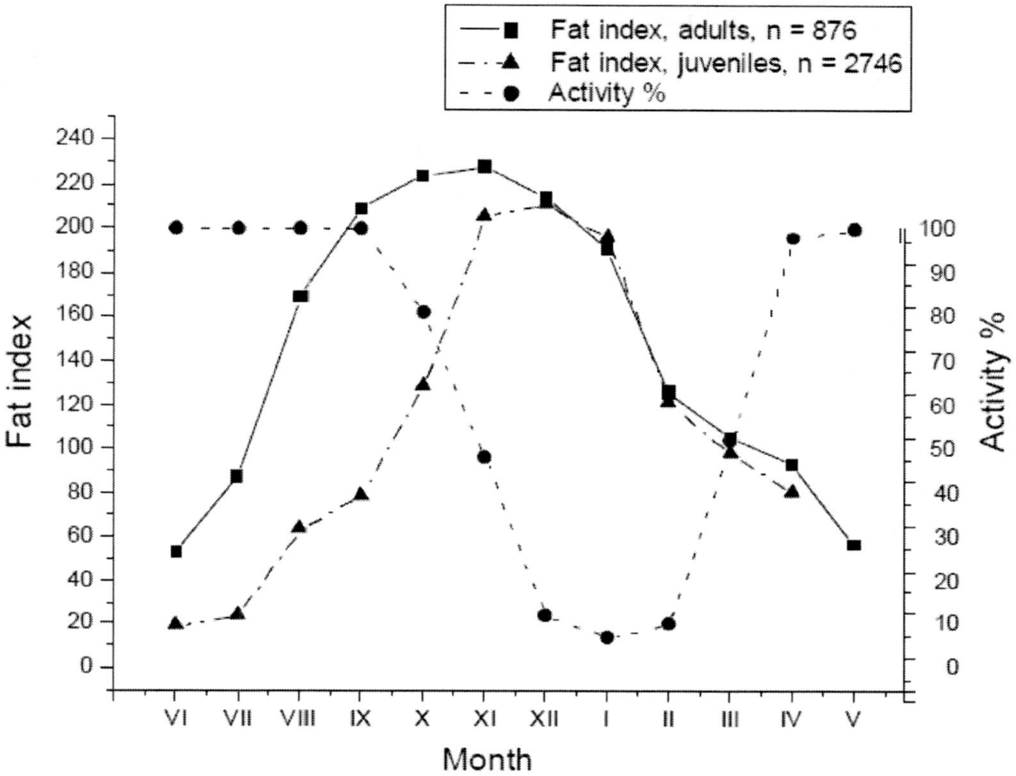

Figure 2. The mean fat indices of adult and juvenile raccoon dogs and activity % in different months. Fat index = 100* (amount of fat, g, around kidneys and at the top of the abdominal cavity)/head-body length (cm). Data are based on raccoon dog carcasses collected from hunters in Finland between 1986 and 1999. N = number of individuals examined. Activity % = percent raccoon dogs active in a given month, according to radio-tracking in SE Finland between 2000 and 2003.

The average body temperature of raccoon dogs is 38 °C in the snow-free season (Mustonen et al. 2007). During hibernation their body temperature decreases 1.4 – 2.1 °C. Circadian variation in body temperature is greater in winter than in summer and body temperature is lowest in the morning (06:00-11:00 a.m.) and highest in the evening (16:00-23:00 p.m.) (Mustonen et al. 2007). In winter, the circadian activity of raccoon dogs is also lowest in the morning and highest around midnight (Kauhala et al. 2007).

In areas where summers are long and winters short juveniles have enough time to fatten themselves before winter lethargy. Further north, such as northern Finland, summers are too short and winters too long for juveniles to accumulate enough fat and survive the winter (Kauhala 1993; Kauhala & Helle 1995). The length of the summer thus determines the northern limit of permanent raccoon dog distribution in Europe, which today lies near the Arctic Circle. However, without winter sleep raccoon dogs could not live so far north they do nowadays, because it would be impossible for them to move in deep snow and to find food in winter.

1.3. High Reproductive Potential

Among canids, raccoon dogs have exceptionally large litters relative to their body weight (Kauhala 1992). In favourable areas, both in the introduced and original ranges, the mean litter size is 8-10 pups (Judin 1977, Helle & Kauhala 1995, Kowalczyk et al. 2009). Maximum recorded litter size in Finland was 16 pups, and maximum number of *corpora lutea* was 23 (Helle & Kauhala 1995). Omnivory, monogamy and winter sleep are probable factors behind the high reproductive potential of raccoon dogs (Kauhala 1996a): some food is always available, males help in pup-rearing and due to winter sleep females are in fairly good condition at the time of ovulation despite of the weather conditions in winter.

Raccoon dogs reach sexual maturity at the age of 10 months (Helle & Kauhala 1995, Kowalczyk et al. 2009). The following figures are based on carcass material of 599 mature females examined in Finland (Helle & Kauhala 1995). Sixty-six percent of one-year old individuals give birth, and the proportion of reproducing females is higher in older females (78% at 2 years, 88% at 3 years, 84% at 4 years and 93% thereafter). The litter size, on the basis of both placental scars and embryos/fetuses, of 1-2-year-old females is lower (mean 8.7) than that of middle-aged (3-5-year old) females (mean 9.5). The litter size declines thereafter, being 8.5 among 6-7-year old females. Although productivity of 1-year-old females is lowest (5.7 pups/female), they produce a large proportion of pups in the population due to their high numbers. The reproductive value of females is highest at the age of two years due to the high mortality rate in the younger age classes (Helle & Kauhala 1993).

Climate affects the productivity of a raccoon dog population. The time of ovulation depends on weather conditions in spring: when snow melts early, pups are born earlier and have more time to grow and accumulate fat reserves before winter than juveniles in the more northern areas (Helle & Kauhala 1995, Kauhala & Helle 1995, Melis et al. 2010). In areas with long winters, raccoon dog females have less fat, produce smaller litters, and a smaller proportion of them reproduce than females in areas with short winters and long summers (Helle & Kauhala 1995; Figure 3). Also food resources affect pup production. Abundance of small rodents affects the fat reserves of raccoon dogs and thus also their litter size (Kauhala & Helle 1995). The crop of berries may even affect the yearly variation in raccoon dog numbers, as it affects the mortality rate of fetuses and probably also that of juveniles during their first autumn and winter (Helle & Kauhala 1991, Helle & Kauhala 1995).

1.4. Monogamy and Parental Care

Radio-tracking studies have shown that raccoon dogs usually are monogamous (e.g., Kauhala et al. 1993b, Drygala et al. 2008a, 2008b). Male and female have a permanent pair-bond which usually lasts for their lifetime. They share their home range and especially in the pup-rearing season, home ranges of neighbouring pairs do not usually overlap. Male and female move and rest together the year round, except in the pup-rearing season in May-June when males spend more time than females at the den guarding and warming the pups (Kauhala et al. 1993b, Kauhala et al. 1998c, Drygala et al. 2008b, Kowalczyk & Zalewski 2011). At daytime males are usually ‘baby-sitting’, whereas females forage. When females nurse the pups they need a lot of energy and spend more time foraging than males. At night they take turns: one is with pups while the other is foraging. During the first month after birth pups are seldom left alone (Ikeda 1983, Yamamoto 1987, Kauhala et al. 1998c).

When pups are 6-8 weeks old, they start to move around the den site. They may follow their parents at night and start foraging themselves. They catch small food items, such as insects, small birds (probably young individuals) and eggs (Kauhala et al. 1998b). In late summer juveniles start to move on their own and most of them disperse during their first autumn (Kauhala et al. 1993b).

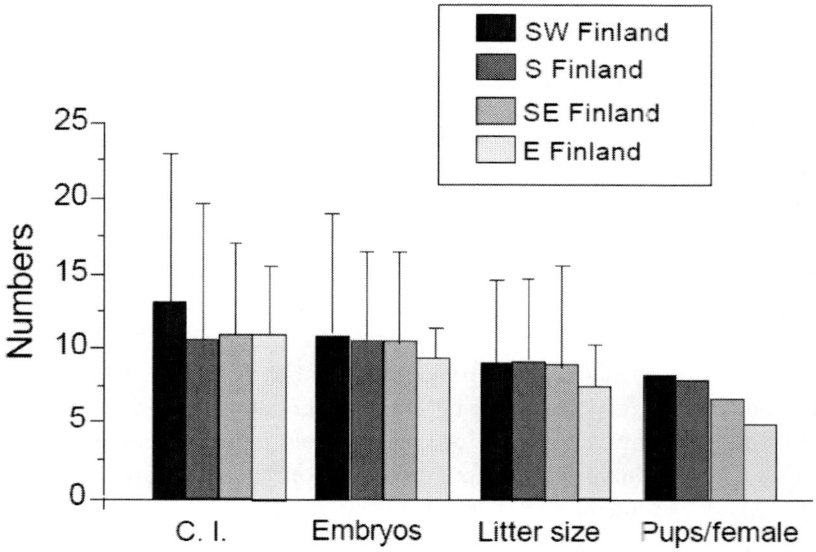

Figure 3. The mean numbers (± SD) of corpora lutea (C. l.), embryos, pups at birth (litter size) and productivity (pups/adult female) of raccoon dog populations in different areas of Finland. Climate is mildest in SW Finland and coldest in E Finland. The mean temperature of the year is approximately 4.7 °C and the length of winter about 130 days in SW Finland, the corresponding figures for the other areas being: 3.8 °C and 140 days for S and SE Finland and 2.4 °C and 175 days for E Finland. Snow depth at the end of March is usually zero in SW Finland, about 5 cm in S Finland, 35 cm in SE Finland and 65 cm in E Finland (Kauhala & Helle 1995, the Finnish Meteorological Institute).

1.5. Tendency to Wander

The mean dispersal distances of juvenile raccoon dogs in southern Finland were estimated to 14-19 km but some individuals may wander as far as 50-70 km, even 145 km, from the marking place (Kauhala et al. 1993b, Kauhala & Helle 1994, Kauhala et al. 2006). Fifty percent of recoveries of ear-tagged juveniles were made within 5 km from the marking place but 21% wandered more than 20 km and 9% more than 40 km (Kauhala et al. 1993b). In northern Germany, the mean dispersal distances of juveniles of both sexes were 13.5 km, and 56% of juveniles were recovered within 5 km from the marking place (Drygala et al. 2010). The maximum distance recorded in northern Germany was 91 km. Both in Finland and in northern Germany most juveniles disperse in September and October but some may disperse as early as in July, and some may not leave their natal home range before late winter

(Kauhala et al. 1993b, Drygala et al. 2010). Dispersal is equal between sexes, probably due to monogamy of the species and the fact that no helpers have been found in raccoon dogs.

On a more or less stable population only young individuals disperse, but adults stay in their home ranges from year to year (Nasimovič & Isakov 1985, Kauhala et al. 1993b, Kauhala & Helle 1994). In a colonizing population also adults may disperse, and dispersal distances are sometimes several hundreds of kilometers (Nasimovič & Isakov 1985, Sutor 2008). Nowak (1973) reported of a dispersal distance of 500 km within three years in a colonizing population. In Sweden, an adult male with a GPS-collar wandered 650 km in a few months (Mikkola 2011).

1.6. Genetic Variation

Genetic variability of the raccoon dog in Europe is probably an effect of combination of different factors including large scale mass introduction, multiple translocations, use of large number of individuals from different geographical areas, and admixture of different mtDNA lineages (Kauhala & Kowalczyk 2011). Phylogenetic analysis indicates different invasion corridors of the species in central Europe (Ansorge et al. 2009) with the secondary contact zone in Germany between different source populations (Pitra et al. 2010). During expansion, some populations were probably created from small number of founders, so population went through several bottlenecks (especially during initial introductions) followed by hybridization of individuals from divergent haplogroups originating from multiple source populations from the native range. Nine haplotypes were found in raccoon dogs in Europe with a sequence divergence of 0.2%–3.2% (mean 1.3%). Haplotype diversity (*h*) was quite high and varied from 0.676-0.746 in populations in Germany to 0.837 in Finland (Pitra et al. 2010).

2. A Vacant Niche? Relationship with Other Medium-Sized Carnivores

2.1. Habitat Use, Home Ranges and Densities

Raccoon dogs prefer damp meadows, marshland and open woodlands where the undergrowth is abundant offering shelter and food for raccoon dogs (Judin 1977, Nasimovič & Isakov 1985, Kauhala & Auttila 2010, Kauhala et al. 2010). They also favour deciduous and mixed forests and fields and often move along lake and river shores (Kauhala 1996b, Holmala & Kauhala 2009). They even escape into water when chased e.g., by a dog (pers. obs.). Raccoon dogs avoid large coniferous forests and other forests with thick canopy and sparse undergrowth (Nasimovič & Isakov 1985, Kauhala & Auttila 2010). However, habitat preferences differ between areas and in northern Germany, no habitat preferences were found (Drygala et al. 2008c).

In many areas they invaded, raccoon dogs became more numerous than native medium-sized carnivores – badgers and red foxes (Kowalczyk et al. 2009). In Finland, the population density of raccoon dogs is highest (up to 2 individuals per 100 ha) in areas which consist of a small scale mosaic of meadows, small deciduous/mixed forests, fields and gardens (Kauhala et al. 2010). Accordingly, home range size of raccoon dogs shows spatial variation. Home ranges (kernel 95%) are smallest (seasonal home ranges about 100 ha) in the most suitable areas but much larger (mean 260 ha in summer) in areas where coniferous forests are common, such as SE Finland (Kauhala et al. 2010). Summer home ranges in northern Germany are almost as large as those in SE Finland (mean males: 194 ha, mean females: 225 ha; Drygala et al. 2008a). In continuous woodlands of the Białowieża Primeval Forest, raccoon dog densities are 0.5-0.7 ind/km^2 (Jędrzejewska & Jędrzejewski 1998).

Raccoon dogs and badgers (*Meles meles*) often share their home ranges but have partly different habitat preferences (Kauhala & Auttila 2010). Although their home ranges largely overlap, the core areas of their ranges tend to be mutually exclusive or overlap only slightly in areas where the most preferred habitats for each species exist (Figure 4). When raccoon dogs prefer damp meadows with abundant undergrowth, badgers favour mature deciduous forests with sparse undergrowth (Kauhala & Auttila 2010). However, when the most preferred habitats for each species do not exist in the area, both species use similar habitats, such as young mixed or deciduous forests and fields (Holmala & Kauhala 2009, Kauhala & Auttila 2010). Both species are thus very flexible in their habitat use, which facilitates the coexistence of these two species.

A radio-tracking study in SE Finland showed, however, that badgers used the preferred habitat in the area (deciduous forest) more often before midnight and raccoon dogs more often after midnight (Kauhala & Holmala 2008). The temporal segregation of habitat use may indicate the need to avoid direct competition between these species.

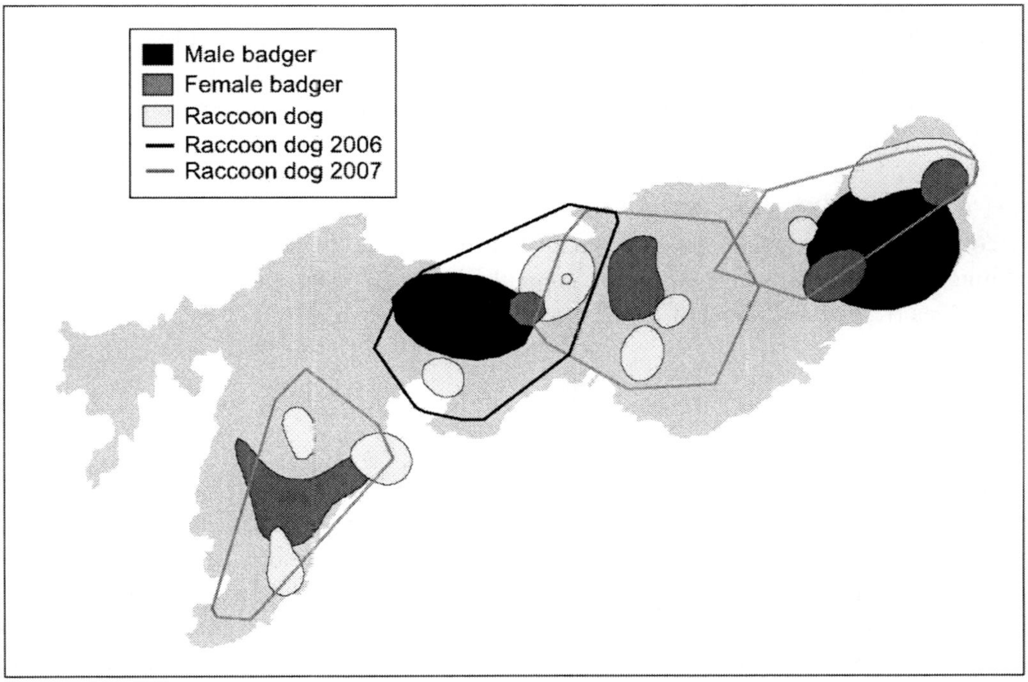

Figure 4. Core areas (kernel 50% home ranges) of raccoon dogs and badgers, and raccoon dog home ranges in an island in SW Finland. Raccoon dogs and badgers were radio-tracked in summers of 2006 and 2007.

Raccoon dogs defecate in a dung pile or latrine, which several individuals may use and which may be used for several years (Ikeda 1984, Yamamoto 1984, Kauhala & Salonen 2012). Many latrines are in the core areas of home ranges but some also lie along the range borders. Latrines are usually located in forests, although raccoon dogs often move also in meadows and field edges (Kauhala & Salonen 2012). Also badgers defecate in certain latrine sites, each of which usually consists of several small pits. In an island in SW Finland, almost every second badger latrine was within a few meters from a raccoon dog latrine (Kauhala & Salonen 2012). Most of these latrines were in oak-hazel groves, which is the preferred habitat for badgers in the area (Kauhala & Auttila 2010). We do not know which species made its latrine in the site first. Badger latrines were, however, situated more often in the core areas of raccoon dogs than *vice versa*. Furthermore, raccoon dog latrines were in permanent places, which may indicate that raccoon dogs were there first and badgers then defecated nearby. These observations point to the conclusion that intra-guild competition for space may exist between raccoon dogs and badgers.

In Finland, the most preferred habitats (mature oak-hazel groves) for badgers are very rare and badger density accordingly much lower than that of raccoon dogs (Kauhala et al. 2006). Badger density is highest in areas with a high proportion of mixed forests and low in areas with vast spruce forests (Kauhala & Holmala 2011).

Fox home ranges are usually larger than those of raccoon dogs, e.g., in SE Finland fox home ranges averaged 660 ha whereas raccoon dog home ranges averaged 390 ha (Kauhala et al. 2006). Fox density in the area was about half (3.5 individuals/10 km^2) of the estimated raccoon dog density (7.7 individuals /10 km^2). Habitat use of raccoon dogs and foxes partly overlapped: both species selected young mixed forests from the landscape (Holmala & Kauhala 2009). Raccoon dogs favoured, however, more often fields than did foxes, whereas foxes preferred more often coniferous forests.

There are very little data of the possible competition between raccoon dogs and foxes. Data collected during a predator removal study in southern Finland suggested, however, that when raccoon dogs were removed from the study area, fox population (snow track index) started to increase (Kauhala 2004). Furthermore, when feces of medium-sized carnivores were collected in SW archipelago of Finland, signs of raccoon dogs and foxes simultaneously inhabiting small islands were not found: in small islands there were either signs of foxes or raccoon dogs (Kauhala & Auniola 2001).

2.2. Shelter Use

Raccoon dogs almost always hide when resting. Most often they rest in dense vegetation (sedges, thickets), under natural shelters created by branches, roots, and trunks of fallen trees, and in different burrows and in hollow trees. Patterns of shelter use may vary according to season, temperature or forest structure (Kowalczyk & Zalewski 2011). To avoid predation and harsh weather conditions raccoon dogs settle in places offering good shelter against predators and harsh weather conditions (Kowalczyk & Zalewski 2011). They hibernate and rear their pups in dens or other shelters. The same dens are often used for several years by the same pairs of raccoon dogs for wintering or reproduction (Kowalczyk & Zalewski 2011). In SE Finland, raccoon dogs used, however, the mean of 3.3 dens during winter (Kauhala et al.

2007). Raccoon dogs may also hide their pups in large dry reed beds (pers. obs.). Reed beds are warmer and there are probably fewer parasites than in burrows.

Raccoon dogs often use badger setts occupying them simultaneously with badgers. Badgers usually start hibernation earlier in autumn and wake up later in spring than raccoon dogs and thus may not be aware of the fact that raccoon dogs accompany them in the den. Sometimes several badgers and raccoon dogs hibernate in the same den and occasionally share the den also in summer (Kauhala & Holmala 2006), but they usually utilize different parts of setts. However, aggressive interactions among the two species were observed, including killing of raccoon dog pups by badgers (Kowalczyk et al. 2008). Because raccoon dogs are not so well adapted to dig dens as are badgers, the occurrence of badgers and their setts in the area may play a facilitative role and help raccoon dogs to colonize the area (Kowalczyk et al. 2008).

2.3. Overlap of Food Niches between Medium-Sized Carnivores

Raccoon dogs, badgers and foxes are all omnivores with a versatile diet. Foxes tend to be, however, more carnivorous than the other two, at least in northern latitudes (Kauhala et al. 1998b). Major part of the diet of foxes in Finland consisted of mammals and birds, but foxes also consumed invertebrates, berries and human refuse (Kauhala et al. 1998b). The food niche of raccoon dogs was wider than that of foxes or badgers consisting of small rodents and shrews, some birds, frogs and lizards, invertebrates, fruits and berries, carrion and waste (Kauhala et al. 1998b, Sutor et al. 2010). The badger also had a versatile diet but consumed more frogs and invertebrates than raccoon dogs (Kauhala et al. 1998b).

In Białowieża Primeval Forest (Poland), food niche overlap between native badgers and red foxes in spring and summer was very low (2%). Food niche overlap was much higher between badgers and raccoon dogs (35%) and highest between foxes and raccoon dogs (41%). In winter, when both foxes and raccoon dogs feed more often on carrion, overlap of their food niche increased to 62% (Jędrzejewska & Jędrzejewski 1998).

The flexible diet of all three species diminishes the possibility of severe competition between them and it seems that they can share the food resources, at least in southern Finland. Furthermore, hibernation of raccoon dogs and badgers in northern areas, such as Finland, results in a situation where there is no competition for food in winter, which is the harshest period of the year. As very efficient scavengers, raccoon dogs may, however, compete with other carnivores for ungulate carcasses (Jędrzejewska & Jędrzejewski 1998, Sidorovich et al. 2000). Sidorovich et al. (2000) pointed out that other carnivores suffer from competition for carcasses in early spring, when other resources are scarce, and their populations declined in Belarus when raccoon dog numbers increased. The polecat (*Mustela putorius*) is one of the species which may have suffered most from the competition for carcasses in winter.

The dentition of Finnish red fox females has slightly changed after the arrival of raccoon dogs in Finland (Viranta & Kauhala 2011). Today the teeth of female foxes show more carnivorous features than before the invasion of the raccoon dog. Vixens are smaller than male foxes, and are thus more likely than males to compete for food with raccoon dogs. This may thus be a case of character displacement in order to avoid competition but also other factors may be involved.

2.4. Changes in Abundance

Estimating raccoon dog population numbers is difficult, because e.g., snow track counts do not give reliable estimates of raccoon dog density due to their winter sleep. We can, however, assume that the change in the hunting bag partly reflects the trend in population numbers. In Finland, hunting bags of the potential competitors of raccoon dogs, i.e., the red fox, badger and also pine marten (*Martes martes*) have not changed much during the past 15 years, possibly indicating that their numbers have not declined, although raccoon dog catch has increased almost three-fold (Figure 5). In Białowieża Primeval Forest, badger densities have not changed after raccoon dog expansion (Kowalczyk et al. 2000). These facts point to the conclusion that there was a vacant niche for the raccoon dog in most of invaded areas. In Belarus, raccoon dogs may have caused a decline of other predators and raptors because of competition for carrion in early spring, when other food resources are scarce (above; Sidorovich et al. 2000).

Although no firm evidence of competition between raccoon dogs and foxes exists, in some areas of Finland, fox numbers tend to decline when raccoon dog population increases rapidly. Furthermore, the fox population is most numerous in SW Finland, whereas the raccoon dog population is most abundant in S and SE Finland (Kauhala 2007).

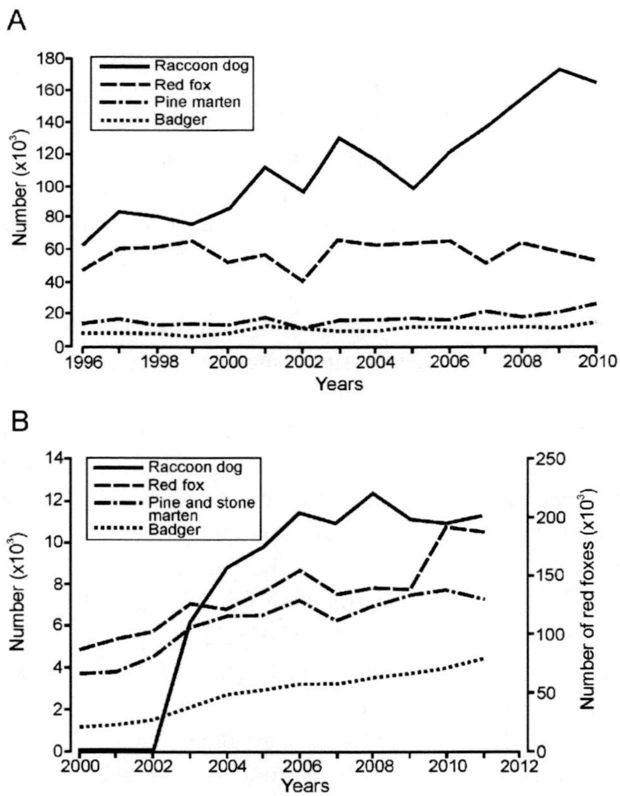

Figure 5. Hunting bags of the raccoon dog, red fox, badger and pine marten in Finland during the past 15 years (Finnish Game and Fisheries Research Institute, A) and in Poland since 2000 (Research Station of Polish Hunting Society, B). Note that the scales are different in the two figures.

2.5. A Vector of Diseases and Parasites

Diseases are an important mortality factor of raccoon dogs (Kowalczyk et al. 2009, Kauhala and Kowalczyk 2011). The raccoon dog's role as a vector of wildlife rabies increased in north-eastern Europe, especially in the Baltic States, during the past few decades (Holmala & Kauhala 2006, WHO 2009). In Europe, there were 148 known cases in raccoon dogs in 1990, whereas the number of known cases was as high as 873 in 2002 (WHO 2002). At the same time period, the number of fox cases decreased from 12,833 to 4,795. In Estonia, there were approximately similar numbers of rabies cases in raccoon dogs and foxes in 2003 but thereafter the number of raccoon dog cases has been higher than that of fox cases (WHO 2003, 2005). Rabid raccoon dogs have also recently been found in Russia, Belarus, Ukraine and Poland (Siemionek et al. 2007, WHO 2010). Furthermore, the raccoon dog was the main vector species (73% of the known cases) during wildlife rabies epizootics in the late 1980s in Finland (Westerling 1991). In Europe, where the fox was the main vector for decades, there is thus a new situation with two important rabies vectors: the raccoon dog and the red fox.

Bait vaccinations have, however, improved the situation, and many European countries are today rabies-free (WHO 2010). For instance, the number of cases has declined sharply in Estonia and Poland during recent years (Figure 6). However, wildlife rabies cases still occur in foxes and raccoon dogs in eastern European countries, most observed cases being in Belarus (544), Croatia (580, not raccoon dogs), Russia (1192) and Ukraine (621; WHO 2010). The total number of rabies cases in wildlife (excluding bats) was 4,310 in Europe in 2010.

Because raccoon dogs share their home ranges and den sites with badgers and foxes and their nightly paths often cross (Kauhala & Holmala 2006), raccoon dogs easily transmit diseases to badgers and foxes. Rabies models constructed for southern Finland showed that an epizootic with two vector species (raccoon dog and red fox) would be stronger than that with only one vector species (Singer et al. 2009). The raccoon dog would be the major host in an area where its density is higher than that of the fox, such as Finland, and the number of fox cases would rather depend on raccoon dog than on fox density. An outbreak of rabies is likely even when the pooled density of the two vector species is lower than the density of each of these species demanded for epizootics with only one vector species (Singer et al. 2009). When the raccoon dog is the main vector species, the risk for an epizootic is highest in autumn when juveniles disperse and in spring when adults move the longest distances (Kauhala et al. 2007). The risk is lowest in winter when raccoon dogs hibernate.

Raccoon dogs may also spread dangerous parasites, such as *Echinococcus multilocularis* (Thiess et al. 2001, Machnicka-Rowińska et al. 2002). The fox has been the major definite host of this parasite (small rodents being the intermediate hosts) in Europe but recently its prevalence in raccoon dogs has increased and is now as high as 6-12% in northern Brandenburg (Germany) and 8% in Poland (Machnicka-Rowińska et al. 2002, Schwarz et al. 2011). The parasite is found in many west and central European countries and is spreading northwards in Europe. It reached Estonia by 2005 and Sweden in 2011 (Moks et al. 2005, Romig et al. 2006, Osterman Lind et al. 2011). The first cases in Estonia and Sweden were found from foxes. It is, however, probable that the parasite came to Sweden with dogs brought from some other country (Osterman Lind et al. 2011).

Echinococcus multilocularis is not dangerous to other canids but it can cause a fatal disease (alveolar echinococcosis) in humans. Raccoon dogs also spread, however, parasites dangerous to other wild carnivores. One of these is *Sarpoctes scabiei* causing sarcoptic mange. This painful skin disease can be an important mortality factor among many carnivores, especially among foxes (Shibata & Kawamichi 1999, Kowalczyk et al. 2009, Kauhala & Kowalczyk 2011). Raccoon dogs are also important reservoirs of *Trichinella* spp. in Finland (Oivanen et al. 2002, Kauhala & Kowalczyk 2011). It carries all fours species of *Trichinella*, which occur in Finland (Oivanen et al. 2002). Furthermore, the prevalence of thrichinosis in wild animals increased in Finland after the arrival of the raccoon dog.

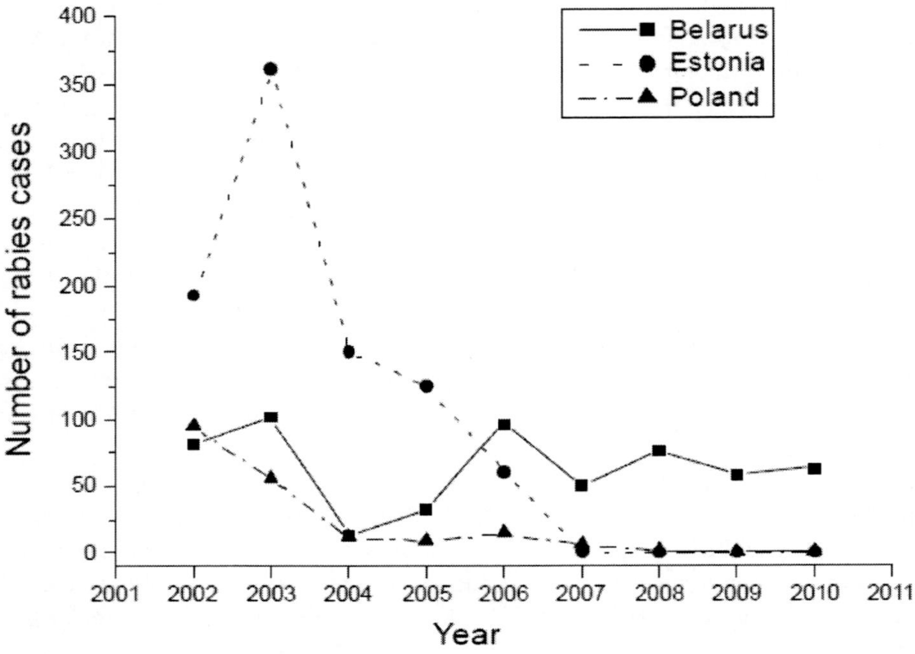

Figure 6. Number of observed rabies cases in raccoon dogs in Belarus, Estonia and Poland in 2002-2010 (WHO).

3. Predation on Native Fauna

Raccoon dogs are suspected to cause much damage to game birds, such as waterfowl and grouse populations, by destroying their nests and preying on juvenile and even on adult birds. However, diet studies do not give much support to this (reviews in Judin 1977, Nasimovič & Isakov 1985, Sutor et al. 2010). Most birds consumed by raccoon dogs are passerines (e.g., Kauhala et al. 1993a, Drygala et al. 2000). In rich habitats of the Białowieża Primeval Forest, birds (mainly jay *Garrulus glandarius*) constituted 13% of the biomass consumed in spring-summer, while only 1.5% in autumn-winter. Grouse and waterfowl occur in most areas only in 0-5% of raccoon dog stomachs/feces (e.g., Naaber 1974, Judin 1977, Danilov et al. 1979, Kauhala et al. 1993a, Ansorge 1998, Kauhala et al. 1998b, Sutor et al. 2010). However, in bird colonies or islands raccoon dogs may prey on birds more often than in mainland areas (Naaber 1971, 1984, Ihalainen 2012).

In SW archipelago of Finland, 67% of raccoon dog feces contained remains of ducks, mainly female eiders (*Somateria mollissima*) but it is likely that the birds had died because of a viral disease, which killed many brooding eiders in the area during the years when the samples were collected (Kauhala & Auniola 2001). Raccoon dogs are known to catch sick or injured birds or consume bird carcasses in other areas, too (e.g., Barbu 1972, Naaber 1974, Viro & Mikkola 1981).

Predator removal studies conducted in Finland do not give much support to the harmfulness of raccoon dogs for game bird populations (Kauhala et al. 2000, Kauhala 2004). In southern Finland, breeding success of ducks correlated positively with raccoon dog abundance both in a predator protection area and a predator removal area where most removed predators were raccoon dogs (Kauhala 2004). In contrast, chick production tended to decrease in the predator removal area at the end of the experiment, together with an increasing fox population. The breeding success of grouse did not correlate with raccoon dog abundance in the study areas, except that of black grouse (*Tetrao tetrix*), which correlated positively with raccoon dog abundance in the predator protection area (Kauhala et al. 2000).

Some pieces of egg shells, including those of domestic poultry, are found in raccoon dog feces but they are usually difficult to identify (Kauhala 2009). In southern Finland, remains of egg shells were found in 11-40% of scats (Kauhala & Auniola 2001). These include egg shells from compost heaps. During a nest predation experiment Opermanis et al. (2001) found that raccoon dogs destroyed only 0.3% of waterfowl nests in a wetland area in Latvia. Predator removal or nest predation studies do thus not give firm evidence that raccoon dogs are harmful to game birds.

Raccoon dogs also commonly prey on frogs (e.g., Barbu 1972, Kauhala & Auniola 2001, Sutor et al. 2010). In SW Finland, frogs were, however, almost absent in the raccoon dog's diet in small islands, which may point to the conclusion that frog populations on the islands are sparse. Frogs were easy prey for alien predators, i.e., raccoon dogs and American minks (*Neovison vison*), and frog populations may have already been destroyed by these.

4. Management of the Population

4.1. Expansion and Distribution

The native distribution range of raccoon dogs in the Far East covers northern Vietnam, Korea, Japan, Amur and Ussuri regions of Siberia and eastern China (Kauhala & Saeki 2004). After mass introduction (over 9,000 individuals released in total) to the European parts of the former Soviet Union between 1929 and 1955, the raccoon dog has spread over eastern, northern and central Europe (Nowak & Pielowski 1964). Pace of spread was around 40 km per year (Lavrov 1971). By the 1980s raccoon dogs had colonized an area over 1.4 million km^2 of Europe (Nowak & Pielowski 1964, Nowak 1973), and still continues its expansion. It reached Finland in 1930s, Sweden in 1945, Baltic countries and Romania in 1950s, Poland in 1955, eastern Germany, Austria and Hungary in early 1960s (Nowak & Pielowski 1964, Kauhala & Kowalczyk 2011).

After first phase of fast expansion and rapid population growth, raccoon dog invasion slowed down. The factors behind this were probably man-changed habitats in western

Europe, less favourable for the species strongly linked to wet natural habitats. In late 1970s raccoon dogs reached France, in 1980s Balkan countries and Norway. In 2000s, raccoon dogs have managed to cross the Alps, and were recorded in northern Italy (P. Genovesi, in litt.). Nowadays, raccoon dogs are found occasionally in Bulgaria, the Netherlands, Macedonia, Switzerland, and even in Spain (Kauhala & Kowalczyk 2011).

4.2. Mortality

Raccoon dog populations are characterized by a very high mortality rate. The annual mortality rate of the population in southern Finland was 81%, being 51-54% for adults and 88-89% for juveniles (Helle & Kauhala 1993). Hunting is a major cause of mortality in Finland. In Białowieża Primeval Forest (Poland) the mortality rate of adults, mainly from predation (from wolves and dogs) and diseases, was 62%, while 57% of juveniles died during the first three months of their life (Kowalczyk et al. 2009). The majority of raccoon dogs (98%) died before reaching the third year of their life (in Finland 1% reached the age of 5 years) and the maximum life span is 7-8 years. Life expectancy at birth is 0.7-0.8 years (Helle & Kauhala 1993, Kowalczyk et al. 2009). High moratlity of juveniles due to collisions with cars is recorded during dispersal (August-September).

4.3. Hunting in Different Countries

The hunting season for raccoon dogs is from 1 August to 30 April in Finland. The annual hunting bag has increased rapidly since the mid-1990s, being today about 160,000 each year (Finnish Game and Fisheries Research Institute; Figure 5 above). About 50% of the autumn population is hunted each year (Kauhala 2007). The size of the autumn population would thus be about 320,000 individuals in Finland, including the juveniles which have survived their first summer. Despite of the high hunting pressure, the hunting bag has increased in recent years. This fact indicates that raccoon dogs are very difficult to eradicate, when the population is established and raccoon dog numbers are high.

The hunting bag has also increased rapidly in Germany since the mid-1990s, being today about 30,000 individuals (Drygala et al. 2008a, 2008b, German Hunter's Union). In Poland, 11,000-12,000 raccoon dogs were hunted annually in 2005-2010 (Data by Research Station of Polish Hunting Society, 2012).

Conclusion

Raccoon dog is one of the invasive carnivores that were introduced to Europe in the 20th century. It has adapted very well to habitats of eastern and northern Europe, and continues its invasion to western and southern parts of the continent (Kauhala & Kowalczyk 2011). Successful expansion of the species was due to a combination of mass introductions into multiple locations, its tendency to wander and the high plasticity and adaptability of raccoon dogs. Its high reproductive capacity enabled fast expansion and population growth, despite of the high mortality rate of raccoon dogs. The high plasticity in adaptation to different climatic and environmental conditions, hibernation and the very wide food niche allowed raccoon dogs' adaptation to local conditions in different areas (Kauhala & Kowalczyk 2011). Finally, low and delayed control efforts in many countries and the secretive way of life were responsible for the unrecorded spread and population establishment in new localities.

Impact of raccoon dogs on native fauna is of minor importance than expected. Locally they may, however, be an important threat to populations of waterfowl and amphibians. Probably the most severe consequence of raccoon dog introduction and expansion is its role as a vector and reservoir of diseases and parasites, such as rabies, sarcoptic mange, trichinellosis and *Echinococcus multilocularis* (Westerling 1991, Oivanen et al. 2002, Oksanen & Lavikainen 2004, Kauhala & Kowalczyk 2011, Kowalczyk & Zalewski 2011).

Raccoon dogs have fitted well in local communities of carnivores and coexist with other medium-sized predators: the red fox and badger. Badgers might facilitate the expansion of raccoon dogs, especially in areas occupied by large predators, through habitat amelioration (setts) and refuge from cold and predation (Kowalczyk et al. 2008).

Wide distribution, locally high densities and continuation of expansion make eradication of raccoon dogs in Europe almost impossible. Raccoon dogs may additionally benefit from climate warming and disperse to areas where cold climates and harsh winters limit today their distribution.

REFERENCES

Ansorge, H. (1998). Biologische Daten des Mardeshundes aus der Oberlausitz. *Abhandlungen und Berichte des Naturkundemuseums Görlitz, Band 70*, 47-61.

Ansorge, H., Ranyuk, M., Kauhala, K., Kowalczyk, R. & Stier, N. (2009). Raccoon dog *Nyctereutes procyonoides* populations in the area of origin and in colonised regions – the epigenetic variability of an immigrant. *Annales Zoologici Fennici 46*, 51-62.

Bannikov, A.G. (1964). [Biology of the raccoon dog in U. S. S. R.] *Mammalia 28*, 1–39 (In French).

Barbu, P. (1972). [Results of studies of raccoon dogs *Nyctereutes procyonoides ussuriensis* Matschie, 1907, from the Danube Delta.] *Säugetierkundliche Mitteilungen 20*, 375-405 (In German).

Danilov, P. I., Rusakov, O. S. & Tumanov, I. P. (1979). [Carnivores of the northwest Soviet Union.] Hiŝnye zveri severo-zabada SSSR. *Nauka, Leningrad*. Pp. 5-25 (In Russian).

Drygala, F., Mix, H. M., Stier, N. & Roth, M. (2000). Preliminary findings from ecological studies of the raccoon dog (*Nyctereutes procyonoides*) in eastern Germany. *Zeitschrift fur Okologie und Naturschutz 9*, 147-152.

Drygala, F., Stier, N., Zoller, H., Bögelsack, K., Mix, H. M. & Roth, M. (2008c). Habitat use of the raccoon dog (*Nyctereutes procyonoides*) in north-eastern Germany. *Zeitschrift für Säugetierkunde 73*, 371-378.

Drygala, F., Stier, N., Zoller, H., Mix, H. M., Bögelsack, K. & Roth, M. (2008a). Spatial organisation and intra-specific relationship of the raccoon dog *Nyctereutes procyonoides* in Central Europe. *Wildlife Biology 14*, 457-466.

Drygala, F., Zoller, H., Stier, N., Mix, H. & Roth, M. (2008b). Ranging and parental care of the raccoon dog *Nyctereutes procyonoides* during pup rearing. *Acta Theriologica 53*, 111-119.

Drygala, F., Zoller, H., Stier, N. & Roth, M. 2010. Dispersal of the raccoon dog *Nyctereutes procyonoides* into a newly invaded area in Central Europe. *Wildlife Biology 16*, 150-161.

Helle, E. & Kauhala, K. (1991). Distribution history and present status of the raccoon dog in Finland. *Holarctic Ecology 14*, 278-286.

Helle, E. & Kauhala, K. (1993). Age structure, mortality, and sex ratio of the raccoon dog in Finland. *Journal of Mammalogy 74*, 936-942.

Helle, E. & Kauhala, K. (1995). Reproduction of the raccoon dog in Finland. *Journal of Mammalogy 76*, 1036-1046.

Holmala, K. & Kauhala, K. (2006). Ecology of wildlife rabies in Europe. *Mammal Review 36*, 17-36.

Holmala, K. & Kauhala, K. (2009). Habitat use of medium-sized carnivores in southeast Finland - Key habitats for rabies spread? *Annales Zoologici Fennici 46,* 233-246.

Ihalainen, A. (2012). [*Diet of raccoon dogs and badgers in Ruissalo Island, Turku.*] – Master thesis, University of Turku, Turku (In Finnish).

Ikeda, H. (1983). Development of young and parental care of the raccoon dog *Nyctereutes procyonoides viverrinus* Temmick, in captivity. *Journal of the Mammalian Society of Japan 9,* 229-236.

Ikeda, H. (1984). Raccoon dog scent marking by scats and its significance in social behaviour. *Journal of Ethology 2*, 77-84.

Ivanova, G. I. (1962). [Comparison of the diets of the red fox, badger and raccoon dog in Voronez national Park.] *Ucenye Zapiski/Moskovskij Gosudarstvennij Pedagogičeskij Institut im. V. I. Lenina 186*, 210-256 (In Russian).

Jędrzejewska, B. & Jędrzejewski, W. (1998). Predation in Vertebrate Communities: the Białowieża Primeval Forest as a case study. *Ecological Studies (Springer Verlag) 135*, 215-219.

Judin, V. G. (1977). [Raccoon dog in Primorje and Priamurye.] *Nauka, Moscow*, Pp. 76-98 (In Russian).

Kauhala, K. (1992). *Ecological characteristics of the raccoon dog in Finland*. Ph.D. thesis, University of Helsinki, Helsinki.

Kauhala, K. (1993). Growth, size, and fat reserves of the raccoon dog in Finland. *Acta Theriologica 38*, 139-150.

Kauhala, K. (1996a). Reproductive strategies of the raccoon dog and the red fox in Finland. *Acta Theriologica 41*, 51-58.

Kauhala, K. (1996b). Habitat use of raccoon dogs, *Nyctereutes procyonoides*, in southern Finland. *Zeitschrift für Säugetierkunde 61,* 269-275.

Kauhala, K. (2004). Removal of medium-sized predators and the breeding success of ducks in Finland. *Folia Zoologica 53*, 367-378.

Kauhala, K. (2007). [The numbers of medium-sized carnivores in Finland.] *Riista- ja kalatalous – Selvityksiä 1*, 1-18 (In Finnish).

Kauhala, K. (2009). [Diet of the omnivorous raccoon dog in Europe and in the Far East.] *Suomen Riista 55*, 45-62 (In Finnish with English summary).

Kauhala, K. & Auniola, M. (2001). Diet of raccoon dogs in summer in the Finnish archipelago. *Ecography 24*, 151-156.

Kauhala, K. & Auttila, M. (2010). Habitat preferences of the native badger and the invasive raccoon dog in southern Finland. *Acta Theriologica* 55: 231-240.

Kauhala K. & Helle, E. (1994). [Home ranges and monogamy of the raccoon dog in southern Finland.] Suomen Riista 40, 32-41 (In Finnish with English summary).

Kauhala, K. & Helle, E. (1995). Population ecology of the raccoon dog in Finland – a synthesis. *Wildlife Biology 1*, 3-9.

Kauhala, K., Helle, P. & Helle, E. (2000). Predator control and the density and reproductive success of grouse populations in Finland. *Ecography 23*, 161-168.

Kauhala, K., Helle, E. & Taskinen, K. (1993b). Home range of the raccoon dog (*Nyctereutes procyonoides*) in southern Finland. *Journal of Zoology (London) 231*, 95-106.

Kauhala, K. & Holmala, K. (2006). Contact rate and risk of rabies spread between medium-sized carnivores in southeast Finland. *Annales Zoologici Fennici 43*, 348-357.

Kauhala, K. & Holmala, K. (2008). Optimal radio-tracking strategy - The best results with the least effort. *Acta Theriologica 53*, 333-341.

Kauhala, K. & Holmala, K. (2011). Landscape features affect home range size of northern badgers (*Meles meles*). *Annales Zoologici Fennici 48*, 221-232.

Kauhala, K., Holmala, K., Lammers, W. & Schregel, J. (2006). Home ranges and densities of medium-sized carnivores in south-east Finland, with special reference to rabies spread. *Acta Theriologica 51*, 1-13.

Kauhala, K., Holmala, K. & Schregel, J. (2007). Seasonal activity patterns and movements of the raccoon dog, a vector of diseases and parasites, in southern Finland. *Mammalian Biology 72*, 342-353.

Kauhala, K., Kaunisto, M. & Helle, E. (1993a). Diet of the raccoon dog, *Nyctereutes procyonoides*, in Finland. *Zeitschrift für Säugetierkunde 58*, 129-136.

Kauhala, K. & Kowalczyk, R. (2011). Invasion of raccoon in dog in Europe: history of introduction, colonization, and threat to the native fauna. *Current Zoology 55*, 584-598.

Kauhala, K., Laukkanen, P. & von Rege, I. (1998b). Summer food composition and food niche overlap of the raccoon dog, red fox and badger in Finland. *Ecogaphy 21*, 457-463.

Kauhala, K., Pietilä, H. & Helle, E. (1998c). Time allocation of male and female raccoon dog to pup rearing at the den. *Acta Theriologica 43*, 301-310.

Kauhala, K. & Saeki, M. (2004). Finnish and Japanese raccoon dogs: On the road to speciation? In: Macdonald, D. W. & Sillero-Zubiro, C. (eds.), *Biology and Conservation of Wild Canids*. Oxford University Press, Oxford, Pp. 217-226.

Kauhala, K. & Salonen, L. (2012). Does a non-invasive method – latrine surveys – reveal habitat preferences of raccoon dogs and badgers? *Mammalian Biology* (in press). Doi: 10.1016/j.mambio.2012.02.007

Kauhala, K., Schregel, J. & Auttila, M. (2010). Habitat impact on raccoon dog *Nyctereutes procyonoides* home range size in southern Finland. *Acta Theriologica 55,* 371-380.

Kauhala, K., Viranta, S., Kishimoto, M., Helle, E. & Obara, I. (1998a). Skull and tooth morphology of Finnish and Japanese raccoon dogs. *Annales Zoologici Fennici 35*, 1-16.

Kobylińska, J. (1996). The red fox and raccoon dog in wetlands of the Biebrza River Valley – Food composition and burrow use. *European Journal of Wildlife Research 1*, 186-189.

Kolosov, A. M., Lavrov, N. P. & Naumov, S. P. (1965). [Biology of game mammals in the Soviet Union.] *Vysšaâ škola, Moscow*, Pp. 69-76 (In Russian).

Korhonen, H. (1987). Energy metabolism of raccoon dog (*Nyctereutes procyonoides*, Gray 1834): applied perspective to common farming practices. *Publications of the University of Kuopio. Original Reports 1/1987,* 1- 69.

Kowalczyk, R., Bunevich, A. N. & Jędrzejewska, B. (2000). Badger density and distribution of setts in Białowieża Primeval Forest (Poland and Belarus) compared to other Eurasian populations. *Acta Theriologica 45*, 395-408.

Kowalczyk, R., Jędrzejewska, B., Zalewski, A. & Jędrzejewski, W. (2008). Facilitative interactions between the Eurasian badger (*Meles meles*), the red fox (*Vulpes vulpes*) and the invasive raccoon dog (*Nyctereutes procyonoides*) in Białowieża Primeval Forest, Poland. *Canadian Journal of Zoology 86*, 1389-1396.

Kowalczyk, R. & Zalewski, A. (2011). Adaptation to cold and predation - shelter use by invasive raccoon dogs *Nyctereutes procyonoides* in Białowieża Primeval Forest (Poland). *European Journal of Wildlife Research 57*, 133-142.

Kowalczyk, R., Zalewski, A., Jędrzejewska, B., Ansorge, H. & Bunevich, A. N. (2009). Reproduction and mortality of invasive raccoon dogs (*Nyctereutes procyonoides*) in the Białowieża Primeval Forest (eastern Poland). *Annales Zoologici Fennici 46*, 291-301.

Lavrov, P. (1971). [Results of raccoon dog introductions in different parts of the Soviet Union.] *Trudy Kafedry Biologii MGZPI 29*, 101-160. (In Russian).

Machnicka-Rowińska, B., Rocki, B., Dziemian, E. & Kołodziej-Sobocińska, M. (2002). [Raccoon dog (*Nyctereutes procyonoides*) - the new host of *Echinococcus multilocularis* in Poland.] *Wiadomosci parazytologiczne 48*, 65-68 (In Polish).

Melis, C., Herfindal, I., Kauhala, K., Andersen, R. & Hogda, K. A. (2010). Predicting animal performance through climate and plant phenology variables: The case of an omnivore hibernating species in Finland. *Mammalian Biology 75*, 151-159.

Mikkola, M. (2011). [A plan for raccoon dog population management.] Suomen riistakeskus, Pohjois-Häme, 23 pp (in Finnish].

Moks, E., Saarma, U. & Valdman, H. (2005). *Echinococcus multilocularis* in Estonia. *Emerging Infectious Diseases 11*, 1973-1974.

Mustonen, A.-M., Asikainen, J., Kauhala, K., Paakkonen, T. & Nieminen, P. (2007). Seasonal rhythms of body temperature in the free-ranging raccoon dog (*Nyctereutes procyonoides*) with special emphasis on winter sleep. *Chronobiology International 24*, 1095-1107.

Naaber, J. (1971). [Raccoon dog.] *Eesti Lodus 14*, 449-455 (In Estonian).

Naaber, J. (1974). [Red fox and raccoon dog in our nature.] In: Merisalu, G. (ed.), *Jaht ja Ulukid. Estonian hunters' handbook 1969-1972*. Eesti Jahimeeste Selts, Tallinn, Pp. 102-115 (In Estonian).

Naaber, J. (1984). [Mammals in Maatsalu today and in the future.] In: Paakspuu V ed. *Eesti NSV Riiklike LooduskaitsealadeTeaduslikud Tööd IV. Maatsalu loodusest.* Valgust, Tallinn (In Estonian).

Nasimovič, A. A. & Isakov, J. A. (eds.) (1985). [Arctic fox, red fox and raccoon dog: distribution of populations, ecology and preservation.] Nauka, Moscow, Pp. 116-145 (In Russian).

Nieminen, P., Mustonen, A.-M., Asikainen, J. & Hyvärinen, H. (2002). Seasonal weight regulation of the raccoon dog (*Nyctereutes procyonoides*): Interactions between melatonin, leptin, ghrelin, and growth hormone. *Journal of Biological Rhythms 17*, 155-163.

Nowak, E. (1973). [Colonization of the raccoon dog (*Nyctereutes procyonoides* Gray) in Europe.] *Beiträge zur Jagd- und Wildforschung 8*, 351-384 (In German).

Nowak, E. & Pielowski, Z. (1964). [Distribution of the raccoon dog in Poland related to its immigration and spread in Europe.] *Acta Theriologica 9*, 81-110 (In German).

Oivanen, L., Kapel, C. M. O., Pozio, E., La Rosa, G., Mikkonen, T. & Sukura, A. (2002). Associations between *Trichinella species* and host species in Finland. *Journal of Parasitology 88*, 84-88.

Oksanen, A. & Lavikainen, A. (2004). Echinococcosis in Fennoscandia. *International Arch. Hylatidosis 35*, 62.

Opermanis, O., Mednis, A. & Bauga, I. (2001). Duck nests and predators: interaction, specialisation and possible management. *Wildlife Biology 7*, 87-96.

Osterman Lind, E., Juremalm, M., Christensson, D., Widgren, S., Hallgren, G., Ågren, E. O., Uhlhorn, H., Lindberg, A., Cedersmyg, M. & Wahlström, H. (2011). First detection of *Echinococcus multilocularis* in Sweden, February to March 2011. *Euro Surveill. 2011 16(14)*:pii=19836. Available online:http://www.eurosurveillance.org/ViewArticle.aspx?ArticleId=19836.

Pitra, C., Schwarz, S. & Fichel, J. (2010). Going west: Invasion genetics of the alien raccoon dog *Nyctereutes procyonoides* in Europe. *European Journal of Wildlife Research 56*, 117-129.

Romig, T., Dinkel, A. & Mackenstedt, U. (2006). The present situation of echinococcosis in Europe. *Parasitology International 55*, S187-S191.

Schwarz, S., Sutor, A., Staubach, C., Mattis, R., Tackmann, K. & Conraths, F. J. (2011). Estimated prevalence of *Echinococcus multilocularis* in raccoon dogs *Nyctereutes procyonoides* in northern Brandenburg, Germany. *Current Zoology 57,* 655-661.

Shibata, F. & Kawamichi, T. (1999). Decline of raccoon dog populations resulting from sarcoptic mange epizootics. *Mammalia 63,* 281-290.

Sidorovich, V. E., Polozov, A. G., Lauzhiel, G. O. & Krasko, D. A. (2000). Dietary overlap among generalist carnivores in relation to the impact of the introduced raccoon dog *Nyctereutes procyonoides* on native predators in northern Belarus. *Zeitschrift für Säugetierkunde 65*, 271-285.

Siemionek, J., Szweda, W. & Radkowski, M. (2007). Epizootic situation of rabies in Warmińsko-Mazurskie Voivodeship in the years 2001-2005. *Medycyna Wweterynaryjna 63*, 700-702.

Singer, A., Kauhala, K., Holmala, K. & Smith, G. C. (2009). Rabies in Northeastern Europe – the threat from invasive raccoon dogs. *Journal of Wildlife Diseases 45*, 1121-1137.

Sutor, A. (2008). Dispersal of the alien raccoon dog *Nyctereutes procyonoides* in Southern Brandenburg, Germany. *European Journal of Wildlife Research 54*, 321-326.

Sutor, A., Kauhala, K. & Ansorge, H. (2010). Diet of the raccoon dog (*Nyctereutes procyonoides*) – a canid with an opportunistic foraging strategy. *Acta Theriologia 55*, 165-176.

Thiess, A., Schuster, R., Nockler, K. & Mix, H. (2001). [Helminth findings in indigenous raccoon dogs *Nyctereutes procyonoides* (Gray, 1843).] *Berl. Münch. Tierärzlt. Wschr. 114*, 273-276 (In German with English summary).

Viranta, S. & Kauhala, K. (2011). Increased carnivory in Finnish red fox females – adaptation to a new competitor? *Annales Zoologici Fennici 48*, 17-28.

Viro, P. & Mikkola, H. (1981). Food composition of the raccoon dog *Nyctereutes procyonoides* Gray, 1834 in Finland. *Zeitschrift für Säugetierkunde 46*, 20-26.

Westerling, B. (1991). [Rabies in Finland and its control 1988-90.] *Suomen Riista 37*, 93-100 (In Finnish with English summary).

WHO (2002). Summarizing the rabies situation in Europe 1990-2002 from the Rabies Bulletin Europe. *Rabies Bulletin Europe 26(4)*, 11-17.

WHO (2003). Distribution of rabies in Europe. *Rabies Bulletin Europe 27(4)*, 8-22.

WHO (2005). Distribution of rabies in Europe. *Rabies Bulletin Europe 29(4),* 7-20.

WHO (2009). Distribution of rabies in Europe. *Rabies Bulletin Europe 33(4),* 9-20.

WHO (2010). Distribution of rabies in Europe. *Rabies Bulletin Europe 34(4),* 12-24.

Yamamoto, I. (1984). Latrine utilization and feces recognition in the raccoon dog, *Nyctereutes procyonoides*. *Journal of Ethology 2*, 47-54.

Yamamoto, I. (1987). Male parental care in the raccoon dog *Nyctereutes procyonoides* during the early rearing period. In: Itô, Y., Brown, J. L. & Kikkawa, J. (eds.), *Animal Societies: Theories and Facts.* Japan Sci. Soc. Press, Tokyo, Pp. 189-195.

In: Advances in Zoology Research. Volume 6
Editor: Owen P. Jenkins
ISBN: 978-1-61122-061-2

Chapter 9

Diversity of Parasitic Nematodes of Marine Fishes from the Gulf of Thailand

Watchariya Purivirojkul*
Animal Systematics and Ecology Speciality Research Unit (ASESRU), Department of Zoology, Faculty of Science, Kasetsart University, Chatuchak, Bangkok, Thailand

Abstract

Eleven genera of parasitic nematodes of marine fishes from the Gulf of Thailand were reported. Five genera of larval Ascaridoid nematodes (*Anisakis, Contracaecum, Terranova, Raphidascaris* and *Porrocaecum*) were found in the mesenteries, visceral organs and intestines of many species of marine fishes. *Anisakis* was found in 4 species, Type I Anisakis larvae were found in 22 species, larvae of the genus *Contracaecum* were found in 2 species, and type A *Contracaecum* larvae were found in 12 species. Type B *Contracaecum* larvae were found in 5 species. Type B *Terranova* larvae were found in 13 species. Larvae of the genus *Raphidascaris* were found in 8 species. Larvae of the genus *Porrocaecum* were found in 2 species. Six genera of mature parasitic nematodes, namely *Capillaria, Cucullanus, Dichelyne* (*Cucullanellus*), *Camallanus, Procamallanus* (subgenus *Spirocamallanus*) and *Philometra*, were also reported in the Gulf of Thailand. *Capillaria* sp. was found in one fish species. *Cucullanus* spp. were found in 7 species, whereas *Dichelyne* (*Cucullanellus*) sp. was found in only one species. Genus *Camallanus* was found in 6 species. *Procamallanus* (subgenus *Spirocamallanus*) was found in 13 species. *Philometra* was found in 11 species. Details regarding parasitic nematode distribution in the Gulf of Thailand were reported, and this distribution was compared with that of another location in this chapter.

Parasites in aquatic animals, especially fish, have been studied by many scientists. Parasitic studies are important because they increase taxonomic knowledge regarding fish populations; additionally, some groups of parasites in marine fish are important to the health of consumers. Many groups of parasites have been reported including protozoa, monogenea, trematodes, cestodes, acanthocephalans, nematodes, copepods and isopods. This chapter is

* Corresponding author: e-mail : fsciwyp@ku.ac.th.

focused on parasitic nematodes, which are invertebrate roundworms. Nematodes are bilaterally symmetrical organisms with a pseudocoelom. These organisms are cylindrical and possess a complete digestive tract with an esophagus, intestine and anus (Williams and Bunkley-Williams, 1996). Most nematodes are dioecious; males are smaller than females and have a curved tail at the posterior (Figure 1), which consists of copulatory structures such as the spicule, bursa and caudal alae.

Nematodes inhabit humid interstitial environments in all habitats (i.e., sea, freshwater and land), and they can also inhabit the bodies of plants and animals (Brusca and Brusca, 1990; Ruppert et al., 2004) .They comprise 256 families and more than 40,000 species (Nagasawa, 2005) .There are many parasitic nematodes displaying all degrees of parasitism that can attack virtually all plants and animals and often form species-specific relationships (Ruppert et al., 2004).

The Gulf of Thailand is a semi-enclosed tropical sea located in the South China Sea (Pacific Ocean); it is surrounded by Malaysia, Thailand, Cambodia and Vietnam (Wattayakorn, 2006). There are many varieties of organisms in the Gulf of Thailand such as phytoplankton, zooplankton, crab, shrimp, shellfish and fish (Boonyapiwat, 1999; Jivaluk, 1999). Vidthayanon (1999) reported 300 species of marine fishes from 18 orders and 89 families in the Gulf of Thailand and the eastern coast of Peninsular Malaysia. Because of the high biodiversity in this area, many parasites were found to be associated with marine aquatic animals. Some parasites require only one host in their life cycle, whereas others require more than one host. For example, some nematodes require an invertebrate or vertebrate host (usually fish) for their larval stage and a fish or marine mammal host for their adult stage (Ogbeibu et al., 2014).

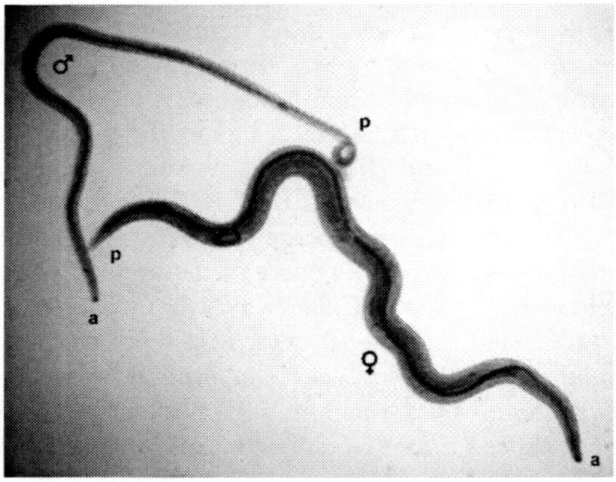

Figure 1. *Procamallanus* sp. from the intestine of *Dendrophysa russelii*. In this image, the size difference between the male and female can be observed (a=anterior, p=posterior).

The numerous species of nematodes that infest humans, food and horticultural crops and domesticated animals make nematodes one of the most important parasitic animal taxa (Ruppert et al., 2004). Most nematodes that parasitize fishes are oviparous, passing eggs into the water that release free-swimming larvae. *Camallanus* and *Philometra* are viviparous (Figure 2) and release larvae directly into the water. The larvae are ingested by an intermediate arthropod host. When the intermediate host is ingested by a fish, the immature nematode either completes development to the adult stage or becomes encysted as a larval form in the tissues or body cavity of the fish. The fish in which larvae encyst may serve as a second intermediate host. The larval life cycle is completed when the fish host is ingested by other carnivorous fish, birds or mammals (Smith and Noga, 1993). However, in many cases, larvae ingested by the wrong fish hosts often survive in the dormant stages in the gut or other tissues for a variable period of time and continue their development into the adult stage if their carrier host is predated by a compatible host; thus, this incompatible fish host can serve as a paratenic host (Petter et al., 1989; Ogbeibu et al., 2014).

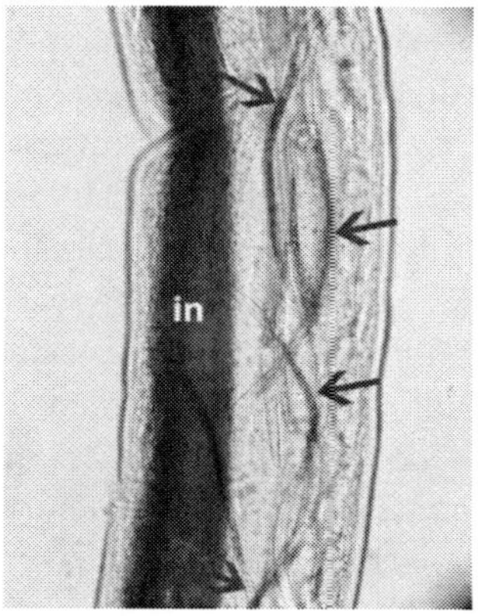

Figure 2. Intestinal region of a *Philometra* sp. parasite of *Lutianus johnii*, showing multiple larvae in the pseudocoelom (arrows). (in=intestine).

Most parasitic nematodes have a heteroxenous life cycle in aquatic animals and must parasitize both an intermediate host and a final host. These nematodes are commonly found in both larval and adult forms in freshwater and marine fishes. Most species of adult nematodes are found in the alimentary canal, with the exception of members of the family Philometridae, which use extraintestinal sites as adults. For example, *Philonema* and *Philometra* parasites are found in the body cavity, liver, eyes, gonads (Figure 3) and musculature of marine fishes (Smith and Noga, 1993; Akhtar, 2008). Additionally, *Anguillicola* parasitizes the swim bladder of eels (Peters and Hartmann, 1986).

Figure 3. Extraintestinal nematodes of marine fishes. A. Ascaridida nematode attached to the liver of *Seriolina nigrofasciata*. B. *Philometra* sp. attached to the ovary of *Lutjanus johnii*.

Nematode larvae are more harmful than adults and can penetrate into the tissues of various organs, causing severe tissue damage and destruction of the cells of the organ (Akhtar, 2008). Parasitic nematodes in the order Ascaridida naturally parasitize fish, cephalopods, marine mammals and piscivorous birds (Purivirojkul, 2009). Members of the genera *Anisakis*, *Pseudoterranova* and *Contracaecum* have similar life cycles (Lymbery et al., 2002). Marine mammals and piscivorous birds serve as definitive hosts for these nematodes, while aquatic invertebrates and fish serve as intermediate hosts. Their larval stage occurs in the internal organs (Figure 3) or the digestive tracts of marine fish. The larvae penetrate the intestine and invade the coelomic cavity or the muscle, where they molt into the third stage and become encapsulated (Bernardi, 2009). *Contracaecum* and *Anisakis* larvae can cause damage to the visceral organs and other tissues during migration through the fish host (Smith and Noga, 1993). The tissue damage endured by the fish leads to a low growth rate and will eventually cause the death of fish with a high level of nematode infestation.

Humans can also become accidental hosts of the Ascaridoidea by ingesting raw marine fish and invertebrates infested with third stage larvae (Anderson, 1992; Szostakowksa et al., 2002; Doupé et al., 2003). In humans, the ingestion of third-stage larvae of some species of the Ascaridoidea can cause a disease known as anisakiasis. In most cases, the nematodes that cause anisakiasis are larvae of *Anisakis simplex* (frequently called type I Anisakis larvae), followed by larvae of *Pseudoterranova decipiens*. Other larval anisakid nematodes, such as *A. physeteris* (type II Anisakis larvae), *Contracaecum* spp., *Contracaecum osculatum* and *Hysterothylacium aduncum*, are very rarely found in humans (Schaum and Müller, 1967; Ishikura, 2003). Since the first discovery of humans infected by *Anisakis* larvae in the Netherlands in 1955, more than 13,000 cases of the disease caused by anisakid worms have been reported from nearly 20 countries until early 1990 (Ishikura et al., 1992).. The areas of highest prevalence are Scandinavia (from cod livers), Japan (from consumption of sushi and sashimi), the Netherlands (from consumption of infected fermented herrings (Maatjes)) and along the Pacific coast of South America (from eating ceviche) (Wikipedia, 2014). Approximately 20,000 cases of anisakiasis were reported in the literature; over 90% of these cases were from Japan, while the rest were from the European Union, United States of America, Canada, New Zealand, Chile and Egypt (Chai et al., 2005; Bernardi, 2009). Anisakiasis occurs when humans ingest raw or lightly cooked marine fish and invertebrates that are infested with anisakis larvae (Nagasawa, 2005). The larvae invade the gastrointestinal mucosa and cause abdominal pain, nausea, vomiting and various gastrointestinal lesions (Kim et al., 2006). The abdominal pain usually occurs after eating raw seafood containing Anisakis larvae. The severity of the pain is di 2 to 7 h fferent among patients (Nagasawa, 2005). Although some species of Ascaridoid nematode do not have any record of causing disease in humans, the presence of Anisakis larvae on and in the viscera and flesh may impact the visual aesthetics and market value of the fish, and parasite removal only adds to the product cost while further reducing its attractiveness for consumers (Doupe′ et al., 2003). Moreover, in some genera, such as *Porrocaecum*, *Raphidascaris* can cause disease in other animals such as fish and birds (Lymbery et al., 2002). Hence, there is a risk in using fish that are infested with larval Ascaridoid nematodes for animal feed (Purivirojkul, 2009).

Parasitic nematodes are harmful and can damage their host if they are present in sufficient numbers. Indeed, parasitic nematodes cause destruction and atrophy of intestinal mucosa as well as necrosis and degeneration of the intestinal tissue (Purivirojkul, 2012). Larval nematodes can cause pathogenesis via their mode of feeding, attachment and movement or migration within the host (Akhtar, 2008). However, these nematodes occasionally have useful roles. For example, some parasites of fish can be used as potential environmental and biological indicators (Landsberg et al., 1998; Palm, 2011; Kleinertz and Palm, 2013) or as biological tags for fishery biologists to help elucidate population movements and other population parameters of fishes (Williams and Bunkley-Williams, 1996). Some nematodes have been used to examine changes in distribution and host ranges caused by anthropogenic activities; these include *Anguillicola crassus*, a parasite in freshwater eels, *Anguilla anguilla* in Europe, *Anguilla rostrata* in North America and *Anguilla japonica* in Taiwan and Japan (Wielgoss et al., 2008). Some species of Ascaridoid nematodes (*Anisakis*) were used as biological tags for host-stock characterization (Mattiucci et al., 2008).

Identification of nematodes is based on external and internal anatomic structures, such as the presence of lips; the shape of the buccal cavity; the length and morphology of the esophagus; the presence of a ventriculus, ventricular appendix or intestinal cecum; the length of the intestinal cecum, the size of spicules, alae and caudal papillae; the location of the vulva; the presence of a copulatory bursa; and the morphology of the tail tip. Nematode identification was performed according to Bykhovskaya–Pavlovskaya et al. (1964), Hoffman (1967), Chabaud (1975), Anderson (1992), Moravec (1994, 2006), Anderson et al. (2009) and Gibbons (2010).

List of parasitic nematodes in marine fishes from the Gulf of Thailand (according to Anderson et al., 2009; Gibbons, 2010)

- Subclass Adenophorea
  - Order Enoplida
    - Superfamily Trichinelloidea
      - Family Trichuridae
        - Subfamily Capillarinae
          - *Capillaria* sp.
- Subclass Secernentea
  - Order Ascaridida
    - Superfamily Ascaridoidea
      - Family Anisakidae
        - Subfamily Anisakinae
          - *Anisakis* spp.
          - *Contracaecum* spp.
          - *Terranova* spp.
        - Subfamily Raphidascaridinae
          - *Raphidascaris* spp.
          - *Raphidascaris* (*Ichthyascaris*) *arii*
      - Family Ascarididae
        - Subfamily Toxocarinae
          - *Porrocaecum* spp.
    - Superfamily Seuratoidea
      - Family Cucullanidae
        - Subfamily Cucullaninae

Cucullanus spp.
Cucullanus rastrelligeri
Cucullanus thaiensis
Dichelyne (*Cucullanellus*) sp.

Order Spirurida
Suborder Camallanina
Superfamily Camallanoidea
Family Camallanidae
Subfamily Camallaninae
Camallanus spp.
Camallanus carangis
Subfamily Procamallaninae
Procamallanus spp.
(subgenus *Spirocamallanus*)
Procamallanus (S.) rigbyi
Procamallanus (S.) similis
Superfamily Dracunculoidea
Family Philometridae
Subfamily Philometrinae
Philometra spp.

The details of each parasitic nematode group found in marine fishes from the Gulf of Thailand are as follows:

SUBCLASS ADENOPHOREA

Adenophoreans do not have a phasmid and possess an excretory system that lacks lateral canals. Caudal papillae are absent or few in number. First stage larvae usually infect the final host (Anderson et al., 2009).

1. Genus *Capillaria*

Capillaria belongs to the superfamily Trichinelloidea, members of which have an esophagus in the form of a stichosome or trophosome with a well-developed intestine and rectum (Anderson et al., 2009). The anterior region of this parasite is more slender than the posterior region, and its esophagus has esophageal glands (stichocytes) (McClelland, 2005). *Capillaria* sp. was classified in the subfamily Capillarinae, family Trichuridae, and the vulva is near the end of the esophagus. Females are oviparous. *Capillaria* sp. has been found in *Gerres oyena* in the Gulf of Thailand (Leethochavalit et al., 1990). *Capillaria* spp. may cause hepatic destruction if the infection is massive, either by virtue of the parasites themselves or their eggs (Ferguson, 1989; Purivirojkul, 2012). *Capillaria* has a wide distribution throughout the African continent; *C. baicalensis* has been reported in *Plectorhinchus macrolepis,* while *C. zederi* has been found in *Pseudotolithus* (*Fonticulus*) *elongates* and *Dasyatis margarita* in

Nigeria (Ogbeibu et al., 2014). In Asia, *C. fudingensis* was found in *Pampus argenteus*, while *Capillaria* sp. was found in *Gerres filamentosus* and *Hexagrammos agrammus* in China (Mamaev, 1970; Wang, 1982; Li et al., 2008; Peng et al., 2011). *C. acanthopagri* has been described as inhabiting the intestine of black porgy, *Acanthopagrus schlegelii schlegelii*, which inhabits the coastal waters of the northwest Pacific Ocean near Japan (Moravec et al., 2010). In India, *C. brevispicula* was found in *Kathala axillaris* and *Nibea albida* (Vijayakumar and Veerappan, 2012). In North America, *C. margolisi* has been described in *Scorpaenichthys marmoratus* near the west coast of Canada (Moravec and McDonald, 2011). In Europe, Ogden (1965) identified *C. wickinsi* in *Pleuronectes platessa* in the southern region of the North Sea.

Taken together, these data indicate that *Capillaria* is highly host-specific (Ogbeibu et al., 2014). This may be the reason for the limited distribution of this parasitic nematode in some host fishes in many marine habitats and also in the Gulf of Thailand.

SUBCLASS SECERNENTEA

Secernenteans have a phasmid and an excretory system with lateral canals. Caudal papillae are almost always numerous in organisms belonging to this subclass. The esophagus does not assume the form of a stichosome. Third stage larvae infect the final host (Anderson et al., 2009). Two orders, Ascaridida and Spirurida, were found in the Gulf of Thailand.

Order Ascaridida

The esophagus of ascarididans is not divided into short muscular and long glandular parts, and these organisms are usually found in the intestine of the final host (Hartwich, 2009). Ascarididans are characterized by the presence of three prominent lips, external labial papillae and numerous caudal papillae (Hartwich, 2009). Members of order Ascaridida, superfamily Ascaridoidea, have a simple esophagus that is cylindrical or terminated by swelling. The esophagus lacks valves and contains uninucleate gland cells. Superfamily Ascaridoidea comprises families Anisakidae (*Anisakis*, *Contracaecum*, *Terranova, Raphidascaris*) and Ascarididae (*Porrocaecum*). These two families were reportedly found in the mesenteries, visceral organs and intestines of marine fish in the Gulf of Thailand (Purivirojkul, 2009).

2. Genus *Anisakis*

Anisakis lack an intestinal caecum and also lack interlabia. The anterior region of the lips forms a medial, slightly bilobed process (Hartwich, 2009). Many species of *Anisakis* in the larval stage were reported in many fishes in multiple locations throughout the world, and these larvae have low host specificity. In the Gulf of Thailand, larvae of the genus *Anisakis* were identified in four species of fish (*Muraenesox* sp., *Epinephelus areolatus, Rachycentron canadum*, *Trichiurus lepturus*) (Purivirojkul, 2009). Type I *Anisakis* larvae were found to parasitize *Apogon* sp., *Selar crumenophthalmus*, *Carangoides gymnostethus*, *Selaroides leptolepis*, *Carangoides malabaricus*, *Decapterus russelli*, *Dussumieria elopsoides*, *Equulites elongatus*, *Lutjanus lutjanus*, *Megalaspis cordyla*, *Nemipterus bipunctatus*, *Priacanthus tayenus*, *Parupeneus cyclostomus*, *Rastrelliger kanagurta*, *Rastrelliger brachysoma*, *Sardinella jussieu*, *Saurida tumbil*, *Scomberomorus commerson*, *Sphyraena jello*, *Terapon jarbua*, *Trichiurus lepturus* and *Upeneus tragula* (Bhaibulaya, 1981; Leethochavalit et al., 1990).

In other locations, this genus is also found in many host species. For example, in China, the larval stage of *A. simplex* was found in 97 species of marine fishes, and the larval stage of *Anisakis* sp. was found in 78 species of marine fishes (Peng et al., 2011). In Korea, Choi et al. (2011) examined 2,537 specimens, consisting of 40 species of fish and 3 species of cephalopods, and found that 34.3% were infected with *A. simplex* third stage larvae (L3). In Yemen, Al-Zubaidy (2010) reported that 20% (94 samples) of 471 marine fishes (5 species belonging to 4 families) were infected with *A. simplex* third stage larvae. In North America, Karl et al. (2011) found *A. simplex* third stage larvae in the muscle and viscera of wild-caught chum salmon (*Oncorhynchus keta*), pink salmon (*O. gorbuscha*) and sockeye salmon (*O. nerka*).

3. Genus *Contracaecum*

Contracaecum exhibits interlabia and an esophagus with a reduced globular ventriculus. The ventriculus has one straight appendix with an intestinal caecum. In the Gulf of Thailand, larvae of the genus *Contracaecum* were found in two species of fish, *Priacanthus tayenus* and *Scolopsis taenioptera* (Purivirojkul, 2009). Moreover, Leethochavalit et al. (1990) reported type A *Contracaecum* larvae in *Eleutheronema tetradactylum, Monacanthus chinensis, Otolithes ruber, Rastrelliger kanagurta, Rastrelliger brachysoma, Gerres oyena, Nemipterus hexodon, Plectorhinchus pictus, Terapon jarbua, Siganus javus, Sphyraena obtusata* and *Scomberomorus guttatus.* Type B *Contracaecum* larvae were found in *Gerres oyena, Nemipterus hexodon, Siganus javus, Rastrelliger brachysoma* and *Rastrelliger kanagurta* (Leethochavalit et al., 1990).

Contracaecum larvae also have low host specificity; they were found in 37 species of marine fishes from China (Peng et al., 2011). These larvae are distributed in many countries. For example, *Contracaecum* larvae were found in *Chorinemus moadetta*, *Otolithes* sp., *Synagris taeniopterus* and *Saurida gracilis* from Borneo (Yamaguti, 1954). *Contracaecum* sp. was found in *Pseudotolithus* (*Fonticulus*) *elongatus* in Nigeria (Ogbeibu et al., 2014). The adult stage of this genus was also found in marine fishes. For example, *C. arii* was found in the small intestine of *Arius* sp. from Borneo, and *Contracaecum* sp. was found in the stomach of *Sphyrna zygaena* from Macassar (Yamaguti, 1954).

4. Genus *Terranova*

Terranova has an intestinal caecum without interlabia. The excretory pore is located between the subventral lips. The anterior region of the lips forms a medial, nearly triangular process that is divided into two indistinct lobes. This genus is usually found as a parasite of elasmobranch fishes (Anderson et al., 2009). In the Gulf of Thailand, type B *Terranova* larvae were found in *Anchoviella* sp., *Carangoides gymnostethus*, *Decapterus russelli, Dussumieria elopsoides, Lutjanus vitta, Megalaspis cordyla, Nemipterus bipunctatus, Rastrelliger kanagurta, Rastrelliger brachysoma, Sardinella jussieu, Siganus javus* and *Trichiurus lepturus* (Bhaibulaya, 1981; Leethochavalit et al., 1990).

Terranova sp. was also found in *Pseudotolithus* (*Fonticulus*) *elongatus* in Nigeria and *Dasyatis akajei* in China (Luo, 2001; Ogbeibu et al., 2014).

5. Genus *Raphidascaris*

Raphidascaris spp. have a ventricular appendix but lack an intestinal caecum. The width of the lips is greater than their length. This organism possesses an excretory pore near the nerve ring. Larvae of *Raphidascaris* were found in 7 species of marine fish from the Gulf of Thailand, consisting of *Dasyatis* sp., *Johnius dussumieri, Nemipterus hexodon, Plotosus lineatus, Pelates quadrilineatus, Rastrelliger brachysoma* and *Terapon jarbua* (Leethochavalit et al., 1990; Purivirojkul, 2009). Recently, *Raphidascaris* (*Ichthyascaris*) *arii* was found in *Arius maculatus* and *Plotosus lineatus* in the Gulf of Thailand (Yooyen et al., 2011c).

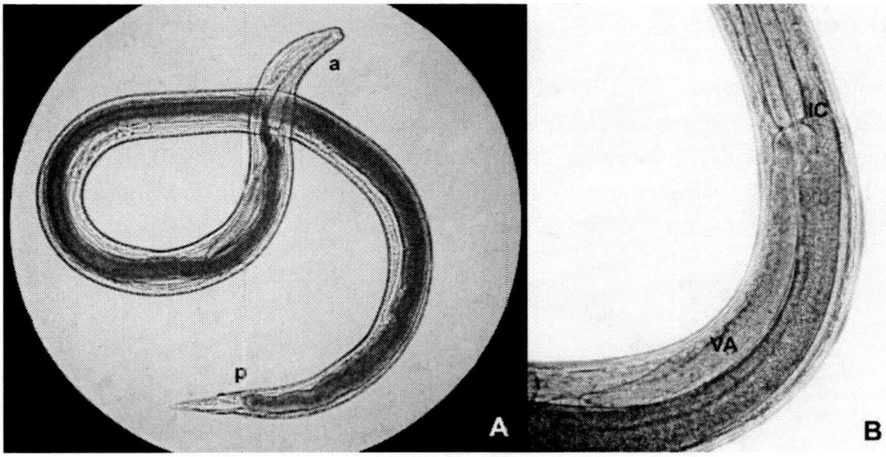

Figure 4. *Contracaecum* spp. from (A). *Scolopsis taenioptera* (a=anterior, p=posterior) and from (B). *Priacanthus tayenus* (IC=intestinal caecum, VA=ventricular appendix).

Raphidascaris spp. have been reported in many parts of the world. In Asia, *Raphidascaris* larvae were found in *Decapterus russelli* from Borneo and in *Echeneis naucrates*, *Nibea albiflora*, *Larimichthys polyactis* and *Lateolabrax japonicus* from China (Yamaguti, 1954; Parukhin, 1967; Ma et al., 2001). Li et al. (2012) described *Raphidascaris* (*Ichthyascaris*) *longispicula* in the intestine and stomach of *Uroconger lepturus* (Anguilliformes: Congridae) from the South China Sea. In North America, *Raphidascaris acus* was reported in *Esox lucius* in Ontario (Smith, 1984). In Africa, *Raphidascaris* sp. was found in *Pseudotolithus* (*Fonticulus*) *elongatus* in Nigeria (Ogbeibu et al., 2014). In the Pacific Ocean, Moravec and Justine (2005) reported *Raphidascaris* (*Ichthyascaris*) *nemipteri* from *Nemipterus furcosus* in New Caledonia. In 2012, Moravec and Justine (2012) reported *Raphidascaris* (*Ichthyascaris*) *etelidis* in the intestine of the *Etelis coruscans* and *Pristipomoides filamentosus* off the southwestern coast of New Caledonia in the southern Pacific Ocean (Moravec and Justine, 2012).

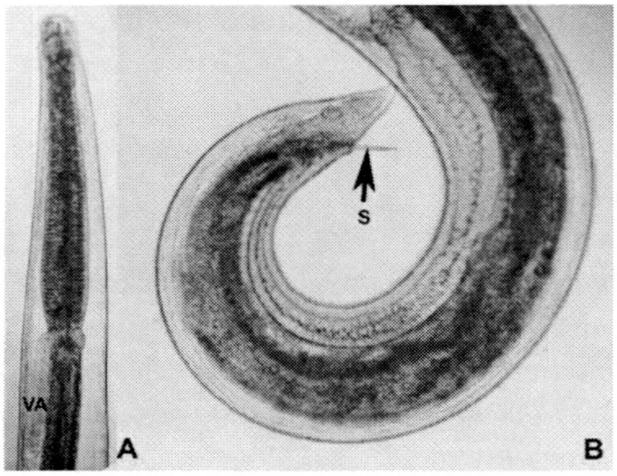

Figure 5. *Raphidascaris* sp. from *Plotosus lineatus*. (A). Anterior region (VA=ventricular appendix). (B). Posterior region (s=spicule).

6. Genus *Porrocaecum*

The anterior margin of each lip of *Porrocaecum* spp. provided an inner surface with a dentigerous ridge. The organism possesses an esophagus with a ventriculus and an intestinal caecum that is as broad or broader than the esophagus (Hartwich, 2009). In the Gulf of Thailand, larvae of *Porrocaecum* were found in two species of fish: *Priacanthus tayenus* and *Seriolina nigrofasciata* (Purivirojkul, 2009).

Figure 6. *Porrocaecum* spp. from *Seriolina nigrofasciata*. (A). Whole nematode (a=anterior, p=posterior). (B). Close-up of the intestinal region (IC=intestinal caecum).

Porrocaecum larvae have low host specificity and are found in 34 species of marine fishes in China (Peng et al., 2011). Yamaguti (1954) found *Porrocaecum* larvae in *Chorinemus moadetta*, *Lethrinus* sp., *Lutianus* sp. and *Megalops cyprinoides* from Borneo. In India, *Porrocaecum reticulatum* was found in *Kathala axillaris* and *Nibea albida*

(Vijayakumar and Veerappan, 2012). The adult stage of *Porrocaecum* sp. was found in the small intestine of *Arius* sp. (Yamaguti, 1954).

7. Genus *Cucullanus*

Cucullanus is a member of the superfamily Seuratoidea, members of which are characterized by a short esophagus that may be divided into two parts with the same or different diameters (Anderson et al., 2009). *Cucullanus* do not have an intestinal caecum or sclerotized plaques in the pseudobuccal capsule and usually have 3 preanal papillae. The esophastome is well developed (Petter, 1974). The identification of species of this genus is based particularly on the length and shape of the spicules, the position of the excretory pore and deirids and the number and arrangement of caudal papillae. *Cucullanus* are parasites of teleosts and amphibians; they occasionally occur in turtles (González-Solís et al., 2007). Most of the approximately 70 species of *Cucullanus* have been described in marine or brackish water fishes of different orders (Hasegawa et al., 1991; González-Solís et al., 2007).

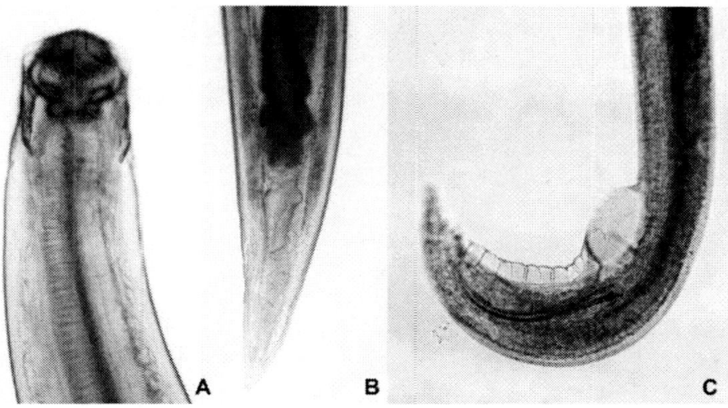

Figure 7. *Camallanus* spp. from *Trichiurus lepturus*. (A). Anterior region of *Camallanus carangis*. (B). Posterior region of a female. (C). Posterior region of a male.

In the Gulf of Thailand, *Cucullanus* sp. was found as a parasite in *Lates calcarifer, Rastrelliger kanagurta, Rastrelliger brachysoma, Terapon jarbua* and *Gerres oyena* (Leethochavalit et al., 1990; Purivirojkul, 2008). Recently, *C. rastrelligeri* and *C. thaiensis* were reportedly found in *Rastrelliger brachysoma* and *Lutjanus vitta*, respectively (Yooyen et al., 2011b). *Cucullanus* sp. was also found in *Lates calcarifer* cultured in Thailand (Chonchuenchob et al., 1987; Ruangpan, 1988). Various species of *Cucullanus* have been found in fishes belonging to different orders such as Perciformes, Pleuronectiformes and Anguilliformes (Petter, 1974; Morand and Rigby, 1998; Moravec et al., 2005; González-Solís et al., 2007). *C. pargi* was found in *Lutjanus griseus* in Mexico, while *Cucullanus* sp. was found in *Pomadasys jubelini* and *Pseudotolithus* (*Pseudotolithus*) *senegalensis* in Nigeria (González-Solís et al., 2007; Ogbeibu et al., 2014). *C. sigani* was found in the small intestine of *Siganus* sp., *C. armatus* was found in the small intestine of *Arius* sp., *C. exiguus* was found in the small intestine of *Lates calcarifer* from Macassar, and *C. arii* was found in the small intestine of *Arius* sp. from Borneo (Yamaguti, 1954). *C. amadai* was found in *Evynnis tumifrons*, *C. hapalogengus* was found in *Hapalogenys mucronatus*, *C. muraenesocis* was found in *Muraenesox cinereus*, and *Cucullanus* sp. was found in *Pleuronichthys cornutus* from China (Luo, 1999; Luo, 2001; Li et al., 2008).

The life cycle of *Cucullanus* has been studied by Køie (2000a, b). Parasitic *C. cirratus* was found in the pyloric caeca and intestine of Atlantic cod, *Gadus morhua*, and may have a life cycle that involves transport hosts (copepods) and intermediate hosts (gobies or cod fry) (Køie, 2000a). *C. heterochrous* uses polychaetes as an intermediate host and various flatfish, including flounder, as final hosts (Køie, 2000b). These intermediate hosts may be the limiting factor for *Cucullanus* distribution.

8. Genus *Dichelyne* (*Cucullanellus*)

Dichelyne is a genus belonging to the family Cucullanidae, subfamily Cucullaninae. This subfamily comprises four genera, including many species that parasitize various freshwater, brackish water and marine fishes worldwide (Caspeta-Mandujano et al., 1999). *Dichelyne* has been separated into three subgenera; however, only *Cucullanellus* or *Dichelyne* (*Cucullanellus*) has been reported in *Gerres oyena* in the Gulf of Thailand (Leethochavalit et al., 1990). *Dichelyne* (*Cucullanellus*) has an intestinal caecum and 11 pairs of caudal papillae, of which 3 pairs are preanal. A preanal sucker is present (Chabaud, 2009).

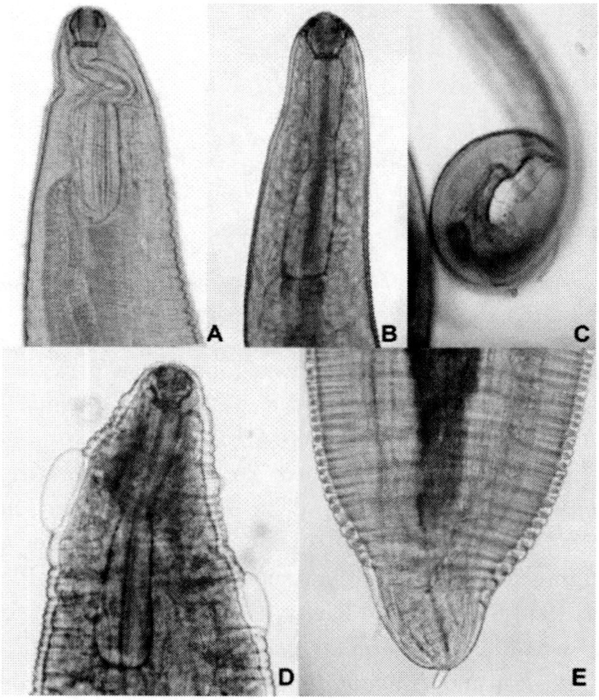

Figure 8. *Procamallanus* spp. (A). Anterior region of *Procamallanus* from *Eubleekeria splendens*. (B). Anterior region of *Procamallanus* from *Dendrophysa russelii*. (C). Posterior region of *Procamallanus* from *Dendrophysa russelii*. (D). Anterior region of *Procamallanus* from *Plectorhinchus diagrammus*. (E). Posterior region of *Procamallanus* from *Plectorhinchus diagrammus*.

Dichelyne is distributed in many parts of the world. Based on the literature, *Dichelyne (Cucullanellus) cotylophora* was found in *Perca flavescens* in Lake Erie, Ontario (Baker, 1984). *Dichelyne (Cucullanellus) pleuronectidis* was found in *Teixeirichthys* sp., and *Dichelyne (Cucullanellus) jialaris* was found in red seabream (*Pagrus major*) in China (Luo, 2001; Luo et al., 2004). *Dichelyne* (*Cucullanellus*) *tornquisti* has been described in the intestine of *Orthopristis ruber* in Rio de Janeiro, Brazil (Paschoal et al., 2014). *Dichelyne* (*Cucullanellus*) *mariajuliae* has been described in *Pogonias cromis* in the Mar Chiquita coastal lagoon, Argentina (Alarcos et al., 2006). *Dichelyne* sp. was found in *Pseudotolithus* (*Fonticulus*) *elongatus* and *Psettias sebae* in Nigeria (Ogbeibu et al., 2014). This genus can found in estuary habitats. For example, the *D. alatae* parasite of *Sillaginopsis panijus* is found in West Bengal, India (De and Maity, 1995).

This genus requires more than two hosts in its life cycle. The life cycle was studied in *Dichelyne* (*Cucullanellus*) *minutus* by Køie (2001). *Dichelyne* (*Cucullanellus*) *minutus* was found in the intestine of *Platichthys flesus, Pleuronectes platessa* and *Pomatoschistus microps* in the Baltic Sea, the Mediterranean Sea and the Black Sea (Gibson, 1972; Fagerholm, 1982; Køie, 2001). The polychaete *Nereis diversicolor* acts as an obligatory intermediate host. *N. diversicolor* may be eaten by the common goby *P. microps*, which harbors this species. The final host, *Platichthys flesus*, receives this parasite either by consuming *N. diversicolor* directly or by consuming infected goby (Køie, 2001).

Order Spirurida

The esophagus of spirurideans is divided into short muscular and long glandular regions. These parasites are found in the anterior region of the gut or in the tissues of final host (Anderson et al., 2009). Two superfamilies Camallanoidea (*Camallanus*, *Procamallanus*) and Dracunculoidea (*Philometra*) have been reported in the Gulf of Thailand. The superfamily Camallanoidea is part of the suborder Camallanina, members of which are parasites of cold-blooded vertebrates and are found in the gut or other organs. Their intermediate hosts are copepods. The esophageal glands are usually uninucleate. The buccal cavity of camallanoideans is well developed, whereas the buccal cavity of superfamily Dracunculoidea is weakly developed (Chabaud, 2009). Dracunculoideans are not usually parasites of the digestive tract; rather, they are usually found in the gonads and occasionally may be found in the eyes of their fish hosts (Chabaud, 2009).

9. Genus *Camallanus*

Camallanus spp. have longitudinal bands sustaining the buccal valves continuous, not separated into ventral and dorsal group (Chabaud, 2009). *Camallanus* spp. were found in six species of fish from the Gulf of Thailand, namely *Nemipterus hexodon, Psettodes erumei*, *Rastrelliger kanagurta, Seriolina nigrofasciata*, *Terapon jarbua* and *Trichiurus lepturus* (Leethochavalit et al., 1990; Purivirojkul, 2008). *Camallanus* sp. found in *Seriolina nigrofasciata* and *Trichiurus lepturus* were classified as *C. carangis*.

Moravec et al. (2006) reported *C. carangis* in *Nemipterus furcosus*, *Upeneus vittatus* and *Parupeneus ciliates*. However, *C. trichuris* was also reportedly found in the fish host

Trichiurus lepturus (Raja Sekhar et al., 2013). This finding indicates that this parasite genus may not be specific to a particular fish host. *C. carangis* was also found in *Caranx* sp. in Fiji, Philippines and in *Caranx* sp., *Scomberoides lysan*, *Selar crumenophthalmus* and *Seriolina nigrofasciata* in China; *C. marinus* and *C. mustelus* were found in *Scomberomorus commerson* and *Mustelus* sp., respectively, in China (Olsen, 1954; Velasquez, 1966; Parukhin, 1976; Wang et al., 1979; Yin, 1983). *Camallanus* sp. was found in *Pseudotolithus* (*Fonticulus*) *elongatus* and *Psettias sebae* in Nigeria (Ogbeibu et al., 2014).

10. Genus *Procamallanus* sp. (subgenus *Spirocamallanus*)

Spirocamallanus was established by Olsen (1952); this genus exhibits spiral thickening on the internal surface of the buccal capsule in both sexes. It is a parasite of fish and amphibians (Gibbons, 2010). Later, Moravec and Sey (1988) considered *Spirocamallanus* to be separated from *Procamallanus* only based on the spiral thickening on the surface of the buccal capsule; thus, *Spirocamallanus* should become a subgenus to accommodate species. However, Chabaud (1975, 2009) and Petter (1979) considered *Spirocamallanus* to be a separate genus. Identification of species belonging to this genus is based particularly on the length and shape of the spicules, position of the excretory pore and deirids and the number and arrangement of the caudal papillae (Agrawal, 1996; González-Solís et al., 2007).

In the Gulf of Thailand, *Procamallanus* (*Spirocamallanus*) sp. was found to parasitize *Gerres oyena, Johnius dussumieri, Lagocephalus spadiceus, Nibea soldado, Otolithes ruber, Plectorhinchus diagrammus, Plectorhinchus pictus, Pomacentrus littoralis, Sciaena* sp., *Scolopsis ciliata*, *Sillago sihama, Sillago maculata*, *Terapon jarbua* (Leethochavalit et al., 1990; Sirikanchana, 2003; Purivirojkul, 2008). *Procamallanus (S.) rigbyi* was found in *Otolithes ruber* and *Procamallanus (S.) similis* was found in *Sillago sihama* (Yooyen et al., 2011a). Based on the literature, *Procamallanus* sp. was found in *Pseudotolithus* (*Fonticulus*) *elongatus* in Nigeria (Ogbeibu et al., 2014). *P. annulatus* was found in the intestine of *Siganus* sp. from Macassar (Yamaguti, 1954). *P. spiralis* was found in the intestine of *Pseudorhombus arsius* from Macassar (Yamaguti, 1954). *Procamallanus (S.) pereirai* was found in *Sillago maculata* and *Sillago sihama* from China (Yin, 1983).

11. Genus *Philometra*

Philometra is a member of the superfamily Dracunculoidea, and the buccal cavity of members of this superfamily is weakly developed. It is not usually found to parasitize the digestive tract. *Philometra* has an elongated, smooth body with more than four cephalic papillae. Females have a blunt tail that is round or slightly tapered. Males have a blunt tail that is rounded and very short; lobes may or may not be present. The organism has a terminal cloaca (Anderson et al., 2009). Based on the characteristics of the cephalic papillae, Chabaud (2009) separated genus *Philometra* into 3 subgenera: *Philometra* (*Ranjhinema*), *Philometra* (*Alinema*) and *Philometra* (*Philometra*).

Philometra spp. was found in 11 species of marine fish in the Gulf of Thailand, namely *Epinephelus areolatus, Epinephelus tauvina, Eubleekeria splendens*, *Equulites elongatus*,

Lobotes surinamensis, *Lutjanus johnii*, *Nibea soldado*, *Platycephalus indicus*, *Psettodes erumei*, *Scolopsis taenioptera* and *Sphyraena jello* (Sirikanchana, 2003; Purivirojkul, 2008).

Philometra spp. are widely distributed in the Atlantic, Indian and Pacific oceans and can be found in brackish water (Moravec, 2006; Moravec et al., 2011). Based on the literature, many species of *Philometra* have been reported in many locations, and their fish hosts are members of the families Lethrinidae, Priacanthidae, Serranidae and Tetraodontidae (Moravec and Justine, 2011). *Philometra* sp. may encyst within the testes, causing trauma when it migrates (Ferguson, 1989). This genus is usually found in the ovaries of many species of marine fishes; for example, *P. brachiri* is found in *Brachirus orientalis*, *P. piscaria* is found in *Epinephelus coioides* and *P. otolithi* is found in *Otolithes ruber* (Moravec and Ali, 2014). *P. brevicollis* and *P. mira* have been described in the ovaries of *Lutjanus vitta* in the southern Pacific Ocean off the coast of New Caledonia (Moravec and Justine, 2011). *Philometra genypteri* has been described in the ovaries of *Genypterus chilensis* was observed in the southeastern Pacific Ocean off the coast of northern Chile (Moravec et al., 2011). The *P. katsuwoni* parasite was found in the gonads of *Katsuwonus pelamis* in the Gulf of Guinea, Brazil (Cárdenas et al., 2009). *P. nemipteri* was found in *Nemipterus virgatus* in China (Luo, 2001). Simmons (1969) found *Philometra* sp. and an unidentified spiruroid infecting the ovaries of approximately 90% of the mature *Katsuwonus pelamis* in the eastern and western tropical Atlantic Ocean. Young fishes lacking mature ovaries were not infected. The highest number of worms found in a pair of ovaries was 75, and there was no obvious damage to the eggs.

In some cases, *Philometra* species, such as *P. tricornuta*, can be found in muscle tissue (Moravec and Ali, 2014). *P. ocularis* was found in the orbital fat tissue of *Epinephelus coioides*, *E. cyanopodus*, *E. rivulatus* and *Variola louti*, while *P. lateolabracis* was found in ovaries of *Epinephelus cyanopodus* and *E. fasciatus* from New Caledonia (Moravec and Justine, 2005). *Philometra* (*Ranjhinema*) *salgadoi* was found in the ocular cavity of the red grouper (*Epinephelus morio*) in the Yucatan Peninsula, Mexico (Vidal-MartÌnez et al., 1995).

A list of parasitic nematodes in marine fishes from the Gulf of Thailand is shown in Table 1.

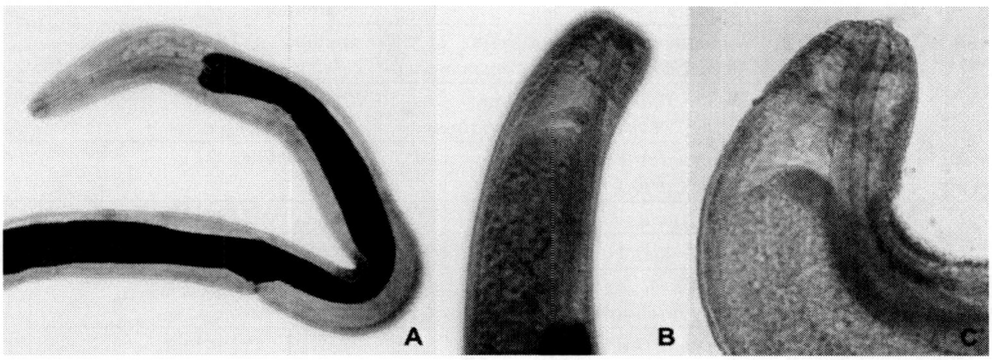

Figure 9. Philometra spp. obtained from (A). *Nibea soldado*, (B). *Lutjanus johnii* and (C). Epinephelus areolatus.

Table 1. A list of parasitic nematodes in marine fishes from the Gulf of Thailand

Genera/Species	Host	References
Capillaria sp.	*Gerres oyena*	Leethochavalit et al. (1990)
Anisakis larva	*Epinephelus areolatus*	Purivirojkul (2008)
	Muraenesox sp.	Purivirojkul (2008)
	Rachycentron canadum	Purivirojkul (2008)
	Trichiurus lepterus	Purivirojkul (2008)
Anisakis larva type I	*Apogon* sp.	Bhaibulaya (1981)
	Carangoides gymnostethus (syn. *Caranx gymnostethoides*)	Bhaibulaya (1981)
	Carangoides malabaricus (syn. *Caranx malabaricus*)	Bhaibulaya (1981)
	Decapterus russelli	Bhaibulaya (1981)
	Dussumieria elopsoides (syn. *Dussumieria hasseltii*)	Bhaibulaya (1981)
	Equulites elongatus (syn. *Leiognathus elongatus*)	Bhaibulaya (1981)
	Lutjanus lutjanus (syn.*Lutianus lineolatus*)	Bhaibulaya (1981)
	Megalaspis cordyla	Bhaibulaya (1981)
	Nemipterus bipunctatus (syn. *Nemipterus delagoae*)	Bhaibulaya (1981)
	Parupeneus cyclostomus (syn. *Pseudupeneus luteus*)	Bhaibulaya (1981)
	Priacanthus tayenus	Bhaibulaya (1981)
	Rastrelliger brachysoma (syn. *Rastrelliger neglectus*)	Leethochavalit et al. (1990)
	Rastrelliger kanagurta	Bhaibulaya (1981)
	Sardinella jussieu	Bhaibulaya (1981)
	Saurida tumbil	Bhaibulaya (1981)
	Scomberomorus commerson (syn. *Scomberomorus commersoni*)	Bhaibulaya (1981)
	Selar crumenophthalmus (syn. *Caranx crumenophthalmus*)	Bhaibulaya (1981)
	Selaroides leptolepis (syn. *Caranx leptolepis*)	Bhaibulaya (1981)
	Sphyraena jello	Bhaibulaya (1981)
	Terapon jarbua	Leethochavalit et al. (1990)
	Trichiurus lepturus (syn. *Trichiurus haumela*)	Bhaibulaya (1981)
	Upeneus tragula	Bhaibulaya (1981)
Contracaecum larva	*Priacanthus tayenus*	Purivirojkul (2008)
	Scolopsis taenioptera	Purivirojkul (2008)
Contracaecum larva type A	*Eleutheronema tetradactylum*	Leethochavalit et al. (1990)
	Gerres oyena	Leethochavalit et al. (1990)

Genera/Species	Host	References
	Monacanthus chinensis	Leethochavalit et al. (1990)
	Nemipterus hexodon	Leethochavalit et al. (1990)
	Otolithes ruber	Leethochavalit et al. (1990)
	Plectorhinchus pictus (syn. *Plectorhynchus pictus*)	Leethochavalit et al. (1990)
	Rastrelliger kanagurta	Leethochavalit et al. (1990)
	Rastrelliger brachysoma (syn. *Rastrelliger neglectus*)	Leethochavalit et al. (1990)
	Scomberomorus guttatus	Leethochavalit et al. (1990)
	Siganus javus	Leethochavalit et al. (1990)
	Sphyraena obtusata	Leethochavalit et al. (1990)
	Terapon jarbua	Leethochavalit et al. (1990)
Contracaecum larva type B	*Gerres oyena*	Leethochavalit et al. (1990)
	Nemipterus hexodon	Leethochavalit et al. (1990)
	Rastrelliger kanagurta	Leethochavalit et al. (1990)
	Rastrelliger brachysoma (syn. *Rastrelliger neglectus*)	Leethochavalit et al. (1990)
	Siganus javus	Leethochavalit et al. (1990)
Terranova larva type B	*Anchoviella* sp.	Bhaibulaya (1981)
	Carangoides gymnostethus (syn. *Caranx gymnostethoides*)	Bhaibulaya (1981)
	Decapterus russelli	Bhaibulaya (1981)
	Dussumieria elopsoides (syn. *Dussumieria hasseltii*)	Bhaibulaya (1981)
	Megalaspis cordyla	Bhaibulaya (1981)
	Lutjanus vitta	Bhaibulaya (1981)
	Nemipterus bipunctatus (syn. *Nemipterus delagoae*)	Bhaibulaya (1981)
	Rastrelliger kanagurta	Leethochavalit et al. (1990)
	Rastrelliger brachysoma	Bhaibulaya (1981)
	(syn. *Rastrelliger neglectus*)	Leethochavalit et al. (1990)
	Sardinella jussieu	Bhaibulaya (1981)
	Selaroides leptolepis (syn. *Caranx leptolepis*)	Bhaibulaya (1981)

Table 1. (Continued)

Genera/Species	Host	References
	Siganus javus	Leethochavalit et al. (1990)
	Trichiurus lepturus (syn. *Trichiurus haumela*)	Bhaibulaya (1981)
Raphidascaris larva	*Dasyatis* sp.	Purivirojkul (2008)
	Johnius dussumieri	Purivirojkul (2008)
	Nemipterus hexodon	Leethochavalit et al. (1990)
	Pelates quadrilineatus	Purivirojkul (2008)
	Plotosus lineatus	Purivirojkul (2008)
	Rastrelliger brachysoma (syn. *Rastrelliger neglectus*)	Leethochavalit et al. (1990)
	Terapon jarbua	Leethochavalit et al. (1990)
Raphidascaris (Ichthyascaris) arii	*Arius maculatus*	Yooyen et al. (2011a)
	Plotosus lineatus	Yooyen et al. (2011a)
Porrocaecum larva	*Priacanthus tayenus*	Purivirojkul (2008)
	Seriolina nigrofasciata	Purivirojkul (2008)
Cucullanus spp.	*Lates calcarifer*	Purivirojkul (2008)
	Gerres oyena	Leethochavalit et al. (1990)
	Rastrelliger kanagurta	Leethochavalit et al. (1990)
	Rastrelliger brachysoma (syn. *Rastrelliger neglectus*)	Leethochavalit et al. (1990)
	Terapon jarbua (syn. *Therapon jarbua*)	Leethochavalit et al. (1990)
Cucullanus rastrelligeri	*Rastrelliger brachysoma*	Yooyen et al. (2011 b)
Cucullanus thaiensis	*Lutjanus vitta*	Yooyen et al. (2011 b)
Dichelyne (*Cucullanellus*) sp.	*Gerres oyena*	Leethochavalit et al. (1990)
Camallanus spp.	*Nemipterus hexodon*	Leethochavalit et al. (1990)
	Psettodes erumei	Purivirojkul (2008)
	Rastrelliger kanagurta	Leethochavalit et al. (1990)
	Terapon jarbua	Leethochavalit et al. (1990)
Camallanus carangis	*Seriolina nigrofasciata*	Purivirojkul (2008)
	Trichiurus lepterus	Purivirojkul (2008)
Procamallanus spp. (subgenus *Spirocamallanus*)	*Gerres oyena*	Leethochavalit et al. (1990)
	Johnius dussumieri	Purivirojkul (2008)
	Lagocephalus spadiceus	Purivirojkul (2008)
	Nibea soldado	Purivirojkul (2008)
	Otolithes ruber	Purivirojkul (2008)

Genera/Species	Host	References
	Plectorhinchus pictus	Leethochavalit et al. (1990)
	(syn. *Plectorhynchus pictus*)	Purivirojkul (2008)
	Plectorhinchus diagrammus	Purivirojkul (2008)
	Pomacentrus littoralis	Sirikanchana (2003)
	Sciaena sp.	Purivirojkul (2008)
	Scolopsis ciliata (syn. *Scolopsis ciliates*)	Purivirojkul (2008)
	Sillago sihama	Sirikanchana (2003)
		Purivirojkul (2008)
	Sillago maculata	Sirikanchana (2003)
	(syn. *Sillago maculate*)	Purivirojkul (2008)
	Terapon jarbua	Leethochavalit et al. (1990)
Procamallanus (S.) rigbyi	*Otolithes ruber*	Yooyen et al. (2011a)
Procamallanus (S.) similis	*Sillago sihama*	Yooyen et al. (2011a)
Philometra spp.	*Epinephelus areolatus*	Purivirojkul (2008)
	Epinephelus tauvina	Sirikanchana (2003)
	Equulites elongatus (syn. *Leiognathus elongatus*)	Sirikanchana (2003)
	Eubleekeria splendens (syn. *Leiognathus splendens*)	Sirikanchana (2003)
	Lobotes surinamensis	Sirikanchana (2003)
	Lutjanus johnii	Purivirojkul (2008)
	Nibea soldado	Purivirojkul (2008)
	Platycephalus indicus	Sirikanchana (2003)
	Psettodes erumei	Purivirojkul (2008)
	Scolopsis taenioptera	Purivirojkul (2008)
	Sphyraena jello	Sirikanchana (2003)

The distribution of parasitic nematodes in the Gulf of Thailand appears to be limited in some cases depending on the life cycle of the parasite. Marine fishes can serve as intermediate hosts, final hosts and/or paratenic hosts. When fish act as the final host, the distribution in host fish species becomes limited. In the Gulf of Thailand, the adult stages of *Capillaria, Cucullanus, Dichelyne* (*Cucullanellus*), *Camallanus, Procamallanus* (subgenus *Spirocamallanus*) and *Philometra* were identified. The copepods are the most common first intermediate host in the life cycle of nematodes in families Camallanidae (*Camallanus, Procamallanus*), Cucullanidae (*Cucullanus, Dichelyne* (*Cucullanellus*)) and Philometridae (*Philometra*) (Ogbeibu et al., 2014); however, in some species, polychaetes can become an intermediate host (Køie, 2000b). Specific marine fish typically served as a final host; however, small fish such as gobies or cod fry sometimes served as a second intermediate host before being consumed by fish.

Compared with parasitic nematodes, which utilize fish as their intermediate host (or paratenic host), these nematodes have low specificity for their host. Especially for nematodes of the ascaridoid group, marine mammals and piscivorous birds serve as definitive hosts, while aquatic invertebrates and fish are intermediate hosts. More than 50 marine fish species were found to harbor the larvae of these nematodes, and many of these fish are paratenic hosts rather than final hosts. Thus, in the Gulf of Thailand and many other areas around the world, this parasitic nematode group can be found in various species of marine fish.

However, only one nematode *Paradujardinia halicoris* was reportedly found in the stomach of the sea cow, *Dugong dugon*, which is a marine mammal in the Gulf of Thailand (Libong island, Trang province). Similar to *Porrocaecum*, a parasite of marine fish in the Gulf of Thailand, *Paradujardinia* is classified in family Ascarididae, subfamily Toxocaridae. *Paradujardinia* has an intestinal caecum and interlabia. Its lips do not have a dentigerous ridge, and its esophagus has a small ventriculus (Hartwich, 2009). The life cycle of this group of nematodes may require an intermediate host (e.g., crustaceans) (Sprent, 1980; Beck and Forrester, 1988), although Jueco (1977) suggested that *Paradujardinia halicoris* has a direct life cycle, as dugongs ingest eggs directly with contaminated sea grasses. However, there are no reports of this parasite in other aquatic animals in the Gulf of Thailand.

REFERENCES

Agrawal, V. (1996). On a New Nematode, *Procamallanus inuelleri* n. sp. from the stomach of a freshwater fish, *Heteropneustes fossilis*. *Proceedings of the Helminthological Society*, *33* (2), 204-208.

Akhtar, Y. (2008). Feeding habits and nematode parasites of some fishes of Karachi coast. Ph.D. thesis. Jinnah university for women for fulfillment.

Alarcos, A. J., Timi, J. T., Etchegoin, J. A. & Sardella. N. H. (2006). A new species of *Dichelyne* (Nematoda: Cucullanidae) parasitizing *Pogonias cromis* (pisces: sciaenidae) from Mar Chiquita coastal lagoon, Argentina. *Journal of Parasitology*, *92*(2), 341–345.

Al-Zubaidy, A. B. (2010). Third-stage larvae of *Anisakis simplex* (Rudolphi, 1809) in the Red Sea fishes, Yemen Coast. *J. KAU. Mar. Sci.*, *21*(1), 95-112.

Anderson, R. C.(1992). *Nematode Parasites of Vertebrates*. Wallingford, CAB International Wallingford.

Anderson, R. C., Chabuad, A. G. & Willmott, S. (2009). Keys to the nematode parasites of vertebrates: archival volumn. CAB International, Wallingford.

Baker, M. R. (1984). Redescription of *Dichelyne* (*Cucullanellus*) *cotylophora* (Ward &Magath, 1917) (Nematoda: Cucullanidae) parasitic in freshwater fishes of eastern North America. *Canadian Journal of Zoology*, *62*(10), 2053-2061.

Beck, C. & Forrester. D. J. (1988). Helminths of Florida manatee, *Trichechus manatus latirostris*, with a discussion and summary of the parasites of sirenians. *Journal of Parasitology*, *74*, 628-637.

Bernardi, C. (2009). Preliminary study on prevalence of larvae of Anisakidae family in European sea bass (*Dicentrarchus labrax*). *Food Control*, 20(4), 433-434.

Bhaibulaya, M. (1981). Ascaridoid nematode larvae in marine fishes from the Gulf of Thailand. *Southeast Asian Journal of Tropical Medicine and Public Health*, *12*(4), 590-594.

Boonyapiwat, S. (1999). Distribution, abundance and species composition of phytoplankton in the South China Sea, Area I: Gulf of Thailand and east coast of Peninsular Malaysia, 111-134. *In* Proceeding of The first technical seminar on marine fishery resources survey in the South China Sea, Area I: Gulf of Thailand and Peninsular Malaysia, 24-26 November 1997, Bangkok, Thailand. Seafdec, Samutprakan, Thailand.

Brusca, R. C. & Brusca, G. J. (1990). Invertebrates. Sinauer Associates, Inc., Massachusetts.

Bykhovskaya – Pavlovskaya, I. E., Guser, A. V., Dubinina, M. N., Izyumora, N. A., Smirnova, I. S., Sokolorskaya, I. L., Shtein, G. A., Shul'man, S. S. & Epsthtein. V. M. (1964). Key to Parasites of Fresh Water Fish of USSR. Translated from Russian. Israel Program for Scientific Translations, Jerusalem.

Cárdenas, M. Q, Moravec, F. & Kohn. A. (2009). First record of *Philometra katsuwoni* (Nematoda, Philometridae), a parasite of skipjack tuna *Katsuwonus pelamis* (Perciformes, Scombridae), off South American Atlantic Coast. *Biota Neotropica*, *9*(2), 263-266.

Caspeta-Mandujano, J. M., Moravec, F. & Salgado-Maldonado, G. (1999). Observations on cucullanid nematodes from freshwater fishes in Mexico, including *Dichelyne mexicanus* sp. n. *Folia Parasitologica*, *46*, 289-295.

Chabaud, A. G. (1975). Keys to the genera of the order Spirurida. Camallanoidea, Dracunculoidea, Gnathostomatoidea, Physalopteroidea, Rictularoidea and Thelazioidea, 1-27. *In* R. C. Anderson, A. G. Chabaud and S. Willmott, eds. CIH Keys to the Nematode Parasites of Vertebrates. No. 3, Part 1. Commonwealth Agricultural Bureaux, Farnham Royal, UK.

Chabaud, A. G. (2009). Spirurida - Camallanoidea, Dracunculoidea, Gnathostomatoidea, Physalopteroidea, Rictularoidea and Thelazioidea, pp. 334-360. *In* R.C. Anderson, A. G. Chabuad and S. Willmott, eds. Keys to the nematode parasites of vertebrates: archival volumn. CAB International, Wallingford.

Chai, J. Y., Murell, K. D. & Lymbery, A. J. (2005). Fish-borne parasitic zoonoses: status and issues. *International Journal for Parasitology*, *35*1233 ,- 1254.

Choi, S. H., Kim, J., Jo, J. O., Cho, M. K., Yu, H. S., Characteristics, H. J. & Ock, M. S. (2011). *Anisakis simplex* Larvae: Infection Status in Marine Fish and Cephalopods Purchased from the Cooperative Fish Market in Busan, Korea. *Korean Journal of Parasitology*, *49*(1), 39-44.

Chonchuenchob, P., Sumpawapol, S. & Mearoh, A. (1987). Diseases of cage-cultured sea bass (*Lates calcarifer*) in southwestern Thailand. ACIAR Proceedings Series, Australian Centre for International Agricultural Research, 20, 194-197.

De, N. C. & Maity, R. N. (1995). A new nematode, *Dichelyne aiatae* sp. n. (Cucullanidae) from *Sillaginopsis panijus* (Pisces) of West Bengal, India. *Folia Parasitologica*, *42*, 220-226.

Doupe', R. G., Lymbery, A. J., Wong, S. & Hobbs. R. P. (2003). Larval anisakid infections of some tropical fish species from north-west Australia. *Journal of Helminthology*, *77*, 363-365.

Fagerholm, H. P. (1982). Parasites of fish in Finland. VI. Nematodes. *Acta Academiae Aboensis Ser B*, *40* (6), 1-128.

Ferguson, H. (1989). Systemic pathology of fish: a text and atlas of comparative tissue responses in diseases of teleosts. Ames: Iowa state university Press.

Gibbons, L. M. (2010). Keys to the nematode parasites of vertebrates: supplementary volumn. *CAB International, Wallingford, UK.*

Gibson, D. I. (1972). Contributions to the life-histories and development of *Cucullanus minutes* Rudolphi, 1819 and *C. heterochrous* Rudolphi, 1802 (Nematoda: Ascaridida). Bulletin of the British Museum (Natural History). *Zoology*, *22* (5), 153-170.

González-Solís, D., Tuz-Paredes, V. M. & Quintal-Loria, M. A. (2007). *Cucullanus pargi* sp. n. (Nematoda: Cucullanidae) from the grey snapper *Lutjanus griseus* off the southern coast of Quintana Roo, Mexico. *Folia Parasitologica*, *54*, 220–224.

Hartwich, G. (2009). Ascaridida-Ascaridoidea, 309-323. *In* R.C. Anderson, A. G. Chabuad and S. Willmott, eds. Keys to the nematode parasites of vertebrates: archival volumn. CAB International, Wallingford.

Hasegawa, H., Williams, E. H. Jr. & Bunkley-Williams, L. (1991). Nematode parasites from marine fishes of Okinawa, Japan. *Journal of the Helminthological Society of Washington*, *58*, 186–197.

Hoffman, G. L. (1967). *Parasites of North America freshwater fishes.* University of California Press, Los Angeles.

Ishikura, H. (2003). *Anisakiasis* (2)*Clinical pathology and epidemiology*, 451- 473. *In* S. Otsuru, S. Kamegai & S. Hayashi, eds. *Progress of Medical Parasitology in Japan*, Vol. *8* .Meguro parasitological museum, Tokyo.

Ishikura, H., Kikuchi, K., Nagasawa, K., Ooiwa, T., Takamiya, H., Sato, N. &

Sugane. K. (1992). Anisakidae and anisakidosis, 43-102. *In* T. Sun, ed. *Progress in clinical parasitology*. *Vol. III*. Springer-Verlag, New York.

Jivaluk, J. (1999). Distribution, abundance and composition of zooplankton in the Gulf of Thailand and East Coast of Peninsular Malaysia, 256-284. *In* Proceedings of the first technical seminar on marine fishery resources survey in the South China Sea, area I. Bangkok.

Jueco, N. (1977). The nematode *Paradujardenia halocoris* in the sea cow in the Philippines. *Philippine Journal of Biology*, *6*, 257-262.

Karl, H., Baumann, F., Ostermeyer, U., Kuhn, T. & Klimpel. S. (2011). *Anisakis simplex* (s.s.) larvae in wild Alaska salmon: no indication of post-mortem migration from viscera into flesh. *Diseases of Aquatic Organisms*, *94*, 201–209.

Kim, Y. H., Choi, W. B., Lee, S. C. and Choi. H. W. (2006). Three cases of

Colonic Anisakiasis. *Korean Journal of Gastrointestinal Endoscopy*, *33*, 239-243. Kleinertz, S. & Palm, H. W. (2013). Parasites of the grouper fish *Epinephelus coioides* (Serranidae) as potential environmental indicators in Indonesian coastal ecosystems. *Journal of Helminthology*, *10*, 1-14.

Køie, M. (2000a). Life cycle and seasonal dynamics of *Cucullanus cirratus* O. F. Müller, 1777 (Nematoda, Ascaridida, Seuratoidea, Cucullanidae) in Atlantic cod *Gadus morhua* L. *Canadian Journal of Zoology*, *78*, 182-190.

Køie, M. (2000b). The life-cycle of the flatfish nematode *Cucullanus heterochrous*. *Journal of Helminthology*, *74*, 323-328.

Køie, M. (2001). The life cycle of Dichelyne (Cucullanellus) minutus Nematoda: Cucullanidae. *Folia Parasitologica*, *48*(4), 304-310.

Landsberg, J. H., Blakesley, B. A., Reese, R. O., Mcrae, G. & Forstchen. P. R. (1998). Parasites of fish as indicators of environmental stress. *Environmental monitoring and assessment*, *51*, 211-232.

Leethochavalit, S., Kliruak, W. & Monvises. A. (1990). A survey of parasites in the economic fishes from the Eastern coast of Thailand. Research reports Srinakharinwirot university. (in Thai)

Li, L., Liu, Y. Y., Liu, B. C., Zhang. L. P. (2012). Morphological and molecular evidence for a new species of the genus *Raphidascaris* (Nematoda: Anisakidae) from marine fishes from the South China Sea. *Parasitology Research*, *110*(4), 1473-1479.

Li, L., Xu, Z. & Zhang. L. P. (2008). A preliminary survey of parasitic nematodes from fishes in the Yellow Sea. *Fisheries Science*, *27*, 283–285.

Luo, D. M. (1999). A study on the infection state of parasitic nematode larva of marine fishes from Taiwan Strait. Journal of Xiamen University *(Natural Science)*, *38*, 604–610.

Luo, D. M. (2001). Notes on nematodes of fishes from Taiwan Strait. II (Nematoda: Spiruridea; Ascarididea). *Acta Zootaxonomica Sinica*, *26*, 281–288.

Luo, D., Guo, S., Fang, W. & Huang. H. (2004). Observations on a cucullanid nematode of marine fishes from Taiwan Strait, *Dichelyne* (*Cucullanellus*) *jialaris* n. sp. *Journal of Parasitology*, *90*(3), 608-611 .

Lymbery, A. J., Doupe', R. G., Munshi, M. A. & Wong. T. (2002). Larvae of *Contracaecum* sp. among inshore fish species of southwestern Australia. *Diseases of Aquatic Organisms*, *51*, 157-159.

Ma, H. W., Jiang, T. J., Quan, F. S., Chen, X. G., Zhang, Y. S., Cui, M. S., Zhi, W. Y. & Jiang. D. C. (2001). Investigation for infection status of anisakid larvae in marine fish, and cephalopods from the Bohai Sea, China. *Journal of Medical Science Yanbian University*, 24, 105–114.

Mamaev, Y. L. (1970). Helminths of some food fishes of the Gulf of Tonkin, 127–190. *In* P. G. Oshmarin, Y. L. Mamaev and B. I. Lebedev, eds. Helminths of animals of south-east Asia. Moscow.

Mattiucci, S., Farina, V., Campbell, N., MacKenzie, K., Ramos, P., Pinto, A.

L., Abaunza, P. & Nascetti. G. (2008). *Anisakis* spp. larvae (Nematoda: Anisakidae) from Atlantic horse mackerel: Their genetic identification and use as biological tags for host stock characterization. *Fisheries Research*, *89*, 146-151.

McClelland, G. (2005). Nematoda (roundworms), 104-115. *In* K. Rohde, ed. Marine parasitology. CSIRO Publishing, Australia.

Morand, S. & Rigby, M. C. (1998). Cucullanin nematodes from coral reef fishes of French Polynesia, with a description of *Cucullanus faliexae* n. sp. (Nematoda: Chitwoodchabaudiidae). *Journal of Parasitology*, *84*, 1213–1217.

Moravec, F. (1994). Parasitic nematodes of freshwater fishes of Europe. Kluwer academic publishers, London.

Moravec, F. (2006). Dracunculoid and anguillicoloid nematodes parasitic in vertebrates. Academia, Prague.

Moravec, F. & Ali. A. H. (2014). Additional observations on *Philometra* spp. (Nematoda: Philometridae) in marine fishes off Iraq, with the description of two new species. *Systematic Parasitology*, *87*(3), 259-271.

Moravec, F. & Justine, J.-L. (2005). Two anisakid nematodes from marine fishes off New Caledonia, including *Raphidascaris* (*Ichthyascaris*) *nemipteri* n. sp. from *Nemipterus furcosus*. *Systematic Parasitology*, *62*(2), 101-110.

Moravec, F. & Justine, J.-L. (2005). Two species of *Philometra* (Nematoda, Philometridae) from serranid fishes off New Caledonia. *Acta Parasitologica*, *50*(4), 323–331.

Moravec, F. & Justine, J.-L. (2011). Two new gonad-infecting *Philometra* species (Nematoda: Philometridae) from the marine fish *Lutjanus vitta* (Perciformes: Lutjanidae) off New Caledonia. *Folia Parasitologica*, *58*(4), 302–310.

Moravec, F. & Justine, J.-L. (2012). *Raphidascaris* (*Ichthyascaris*) *etelidis* n. sp. (Nematoda, Anisakidae), a new ascaridoid nematode from lutjanid fishes off New Caledonia. *Zoosystema*, *34*(1),113-121.

Moravec, F., Justine, J.-L. & Rigby. M. C. (2006). Some camallanid nematodes from marine perciform fishes off New Caledonia. *Folia Parasitologica*, *53*, 223–239.

Moravec, F. & Sey. O. (1988). Nematodes of freshwater fishes from North Vietnam. Part 1. Camallanoidea and Habronematoidea. *Vestni^acutek Československé Společnosti Zoologické*, *52*(2), 12-148.

Moravec, F. & McDonald. T. E. (2011). *Capillaria margolisi* sp. nov. (Nematoda: Capillariidae) from a marine fish, *Scorpaenichthys marmoratus* (Ayres), from the west coast of Canada. *Canadian Journal of Zoology*, *59*(1), 88-91.

Moravec, F., Nagasawa, K. & Madinabeitia. I. (2010). A new species of *Capillaria* (Nematoda: Capillariidae) from the intestine of the marine fish *Acanthopagrus schlegelii schlegelii* (Sparidae) from Japan. *Journal of Parasitology*, *96*(4), 771-774.

Moravec, F., Sasal, P., Würtz, J. & Taraschewski. H. (2005). *Cucullanus oceaniensis* sp. n. (Nematoda: Cucullanidae) from Pacific eels (*Anguilla* spp.). *Folia Parasitology*, *52*, 343–348.

Moravec, F., Chávez, R. A. & Oliva. M. E. (2011). A new gonad-infecting species of *Philometra* (Nematoda: Philometridae) from the red cusk-eel *Genypterus chilensis* (Osteichthyes: Ophidiidae) off Chile. *Parasitology Research*, *108*, 227–232.

Nagasawa, K. (2005). Anisakiasis, 430-434. *In* K. Rohde, ed. Marine Parasitology. CSIRO Publishing, Australia.

Ogbeibu, A. E., Okaka, C. E. & Oribhabor. B. J. (2014). Gastrointestinal helminth parasites community of fish species in a Niger Delta Tidal Creek, Nigeria. *Journal of Ecosystems*. http://dx.doi.org/10.1155/ 2014/246283.

Ogden, C. G. (1965). A new species of *Capillaria* (Nematoda), *C. wickinsi* sp. nov., from the marine fish, *Pleuronectes platessa*. *Annals and Magazine of Natural History: Series*, *138*(92), 451-453.

Olsen, L. S. (1954). A new species of *Camallanus* (Nematoda) from a Fijian marine fish. *Transactions of the American Microscopical Society*, *73*, 258-260.

Palm, H. W. (2011). Chapter 12 Fish parasites as biological indicators in a changing world: Can we monitor environmental impact and climate change?, 223-250. *In* H. Mehlhorn, ed. Progress in parasitology, *Parasitology Research* Monographs 2. Springer-Verlag Berlin Heidelberg.

Parukhin, A. M. (1967). On the helminth fauna of the fish Echeneis naucrates from the South China Sea. Uchenye Zapiski Gorkovskogo Gosudarstvvennyi Pedogogicheskii Instituta imeni M. Gorkogo, *Seriya Biologicheskikh Nauk*, *66*, 24–32.

Parukhin, A. M. (1976). Parasitic worms of food fishes from the southern seas. Kiev, Naukova Dumka.

Paschoal, F., Vieira, F., Cezar, A. & Luque. J. (2014). *Dichelyne* (*Cucullanellus*) *tornquisti* n. sp. (Nematoda, Cucullanidae) from corocoro grunt, *Orthopristis ruber* (Cuvier, 1830) (Perciformes, Haemulidae) from southeastern brazil. *Journal of Parasitology*, In-Press.doi: http://dx.doi. org/10.1645/13-334. 1.

Peng, W. F., Liu, S. F., Wang, B. L. & Wei. M. M. (2011). A checklist of parasitic nematodes from marine fishes of China. *Systematic Parasitology*, *79*, 17-40.

Peters, G. & Hartmann. F. (1986). *Anguillicola*, a parasitic nematode of the swim bladder spreading among eel populations in Europe. *Diseases of Aquatic Organisms*, *1*, 229-230.

Petter, A. J. (1974). Essai de classification de la famille des Cucullanidae. Bulletin du Musee d'Histoire Naturelle de Paris, series 3, *Zoologie*, *177*, 1469–1491.

Petter, A. J. (1979). Essai de classification de la sous-famille des Procamallaninae (Nematoda, Camallanidae). *Bulletin du Museum National d'Histoire Naturelle Section A Zoologie Biologie et Ecologie Animales*, *1*, 219-239.

Petter, A. J., Fontaine, Y. A. & Le Belle, N. (1989). Edute du development larvaire de *Anguinicola crassus* (Darcunloidea, Nematoda) chez un cyclopi de la region parisienne. *Annales de Parasitologie Humaine et Comparee*, *64*, 344–355.

Purivirojkul, W. (2008). Biodiversity and distribution of parasites and microoganisms from marine fishes in Chon Buri province. Final report. Kasetsart University Research and Development Institute, Kasetsart University, Bangkok. (in Thai)

Purivirojkul, W. (2009). An investigation of larval Ascaridoid nematodes in some marine fish from the Gulf of Thailand. *Kasetsart Journal (Natural Science)*, *43*(5), 85-92.

Purivirojkul, W. (2012). Histological Change of Aquatic Animals by Parasitic Infection, 153-176. *In* Enrique Poblet Martinez. Histopathology-Reviews and Recent Advances. Intech, DOI: 10.5772/52769. ISBN 978-953-51-0866-5.

Raja Sekhar, M., Baby Ratna Kumari, T. & Sarat Babu, E. (2013). A Study on *Camallanus trichiuris* Bashirullah and Rahman, 1972 (Nematoda: Camallanidae) of two marine fishes, *Trichiurus lepturus* and *Lepturacanthus savala* from Visakhapatnam Coast Andhra Pradesh, India. *International Journal of Biology, Pharmacy and Allied Sciences*, *2*(12), 2443-2448.

Ruangpan, L. (1988). Chapter X diseases of cultured seabass, *Lates calcarifer*. In training manual 88/3 seabass (*Lates calcarifer*) culture in Thailand .The UNDP/FAO regional seafarming project in cooperation with the UNDP/FAO/ASEAN Small-Scale Coastal Fisheries Development Project. http://www. fao. org/docrep/field/003/ab707e/ab707e10. htm.

Ruppert, E. E., Fox, R. S. & Barnes. R. D. (2004). *Invertebrate Zoology*. 7th ed. Thomson Brooks/Cole, United States. Schaum, E. & Müller. W. (1967). Die Heterocheilidiasis. Eine Infection des Menschen mit Larven Von Fisch-Ascariden. *Deutsche Medizinishe Wochenschrift*, *92*, 2230-2233.

Simmons, D. C. (1969). Maturity and spawning of skipjack tuna (*Katsuwonus pelamis*) in the Atlantic Ocean, with comments on nematode infestation of the ovaries. U. S. Fish and Wildlife Service, *Special Scientific Report Fisheries*, *580*, 1-17.

Sirikanchana, P. (2003). *Parasites of Aquatic Animal*. 6th edition. Sky Word Advertising Ltd., Bangkok. (in Thai)

Smith, J. D. (1984). Taxonomy of *Raphidascaris* spp. (Nematoda, Anisakidae) of fishes, with a redescription of *R. acus* (Bloch, 1772). *Canadian Journal of Zoology*, *62*(4), 685-694.

Smith, S. A. and Noga. E. (1993). *General Parasitology*, 132-148. *In* M. K. Stoskopf, ed. Fish medicine. W. B. Saunders company, London.

Sprent, J. F. A. (1980). Ascaridoid nematodes of Sirenians-the Heterocheilinae redefined. *Journal of Helminthology*, *54*, 309-327.

Szostakowska, B., Myjak, P. & Kur, J. (2002). Identification of anisakid nematodes from the southern Baltic Sea using PCR-based methods. *Molecular and Cellular Probes*, *16*, 111-118.

Velasquez, C. C. (1966). Some parasitic helminthes of Philippine fishes. *University of the Philippines Research Digest*, *5*, 23-29.

Vidal-MartÌnez, V. M., Aguirre-Macedo, M. L. & Moravec. F. (1995). *Philometra* (*Ranjhinema*) *salgadoi* n. sp. (Nematoda: Philometridae) from the ocular cavity of the red grouper *Epinephelus morio* (Pisces: Serranidae) from the Yucatan Peninsula, Mexico. *Journal of Parasitology*, *81*, 763-766.

Vidthayanon, C. (1999). Species composition and Diversity of Fishes in the South China Sea, Area I: Gulf of Thailand and East Coast of Peninsular Malaysia, 172-240. *In Proceedings of the First Technical Seminar on Marine Fishery Resources Survey in the South China Sea*, Area I: Gulf of Thailand and East coast of peninsular Malaysia. Southeast Asian Fisheries Development Center: Bangkok, Thailand. Vijayakumar, P. & Veerappan. N. (2012). Endohelminth parasites of two scianidae fishes from the south-east coast of India. *International Journal of Environmental Biology*, *2*(3), 183-186.

Waikagul, J., Thonghom, S. & Maipanich. W. (1998). *Paradujardinia halicoris* Owen, 1833, a nematode from the stomach of a sea cow in the Gulf of Thailand. *Thai Journal of Veterinary Medicine*, *28*(3), 49-56.

Wang, P. Q. (1982). Studies on nematodes of the Family Capillariidae from Fujian. *Acta Zootaxonomica Sinica*, *7*, 126–177.

Wang, P. Q., Zhao, Y. R., Wang, X. Y. & Zhang. J. Y. (1979). Report on some nematodes from vertebrate animals in Central, & South China. *Journal of Fujian Teachers University (Natural Science)*, *2*, 78–92.

Wattayakorn, G. (2006). Environmental Issues in the Gulf of Thailand, 249-259. *In* E. Wolanski, ed. *The Environment in Asia Pacific Harbours*. Springer Netherlands.

Wielgoss, S., Taraschewski, H., Meyer, A. & Wirth, T. (2008). Population structure of the parasitic nematode *Anguillicola crassus*, an invader of declining North Atlantic eel stocks. *Molecular Ecology*, *17*(15), 3478-3495.

Williams, E. H. Jr. & Bunkley-Williams, L. (1996). Parasites of offshore big game fishes of Puerto Rico and the western Atlantic. Puerto Rico Department of Natural and Environmental Resources, San Juan, PR and the University of Puerto Rico, Mayaguez, PR .Wikipedia. (2014). Anisakis. Available Source: http://www.wikipedia.com, 27 March 2014.Yamaguti, S. 1954. Parasitic worms mainly from Celebes. Part 9. Nematodes of Fishes. *Acta Medica Okayama*, *9*(1), 122-133.

Yin, W. Z. (1983). On some nematodes from marine fishes in Hainan Island, China. *Acta Zootaxonomica Sinica*, *8*, 225–228.

Yooyen, T., Moravec, F. & Wongsawad. C. (2011a). Two new sibling species of *Procamallanus* (*Spirocamallanus*) (nematoda: camallanidae) from marine fishes in the gulf of Thailand. *Journal of Parasitology*, *97*(5), 931-938.

Yooyen, T., Moravec, F. & Wongsawad. C. (2011b). Two new species of *Cucullanus* Müller, 1777 (Nematoda: Cucullanidae) from marine fishes off Thailand. *Systematic Parasitology*, *78*(2), 139-149.

Yooyen, T., Moravec, F. & Wongsawad. C. (2011c). *Raphidascaris* (*Ichthyascaris*) *arii* sp. n. (Nematoda: Anisakidae), a new ascaridoid nematode from marine catfishes in the Gulf of Thailand. *Helminthologia*, *48*(4), 262-267.

In: Advances in Zoology Research. Volume 6
Editor: Owen P. Jenkins
ISBN: 978-1-61122-061-2

Chapter 10

THE ROLE OF SNAILS IN RENAL EPITHELIAL TO MESENCHYMAL TRANSITION

S. Saad and C. A. Pollock
Kolling Institute of Medical Research, University of Sydney, Australia

ABSTRACT

Epithelial-mesenchymal transition (EMT) is a highly controlled process involved in organogenesis, that is increasingly recognized as an integral part of tissue repair, organ fibrosis and cancer progression. *In vitro* data have shown that EMT is regulated by a complex signalling network. Recent *in vivo* data, in particular using tracing technologies, have conclusively demonstrated EMT in fibrotic kidney disease, and confirmed that EMT contributes to the pool of interstitial matrix-producing fibroblasts. In this review, we will summarize the known signalling pathways regulating EMT in renal proximal tubule cells following divergent insults, including exposure to high glucose, transforming growth factor β (TGF-β), angiotensin II and hypoxic conditions. We will also describe the signalling pathways/mediators differentially involved in the regulation of EMT in the fibrotic kidney, including Notch, Smad, integrin-linked kinase, Wnt/β-catenin and Erk1/2 and their convergent effects on Snail. The potential role of Snail as a common 'master switch' integrating various signal inputs in controlling EMT, its role in different models of renal pathology and potential therapeutic implications will be discussed.

INTRODUCTION

Epithelial-to-mesenchymal transition (EMT) is a process where differentiated epithelial cells lose their phenotype and acquire characteristic features of mesenchymal cells. It is a highly orchestrated regulated process, resulting in the loss of epithelial cell adhesion, de novo α-smooth muscle actin expression and cytoskeletal reorganization, disruption of tubular basement membrane, and lastly enhanced cell migration and invasion [1]. This process was initially recognized during embryonic development and has more recently been implicated in tumour progression and fibrosis [2, 3].

EMT has been recognised as an essential mechanism involved in tubulointerstitial fibrosis. Tubulointerstitial fibrosis is a common final pathway leading to progressive renal injury in diverse disease. *In vitro* and *in vivo* studies have demonstrated that EMT is a complex process that is initiated by multiple insults and is regulated by many growth factors and cytokines. In this review, we will discuss whether these pathways converge on a specific EMT regulator or a "master switch" that is able to control EMT and hence provide a potential therapeutic target for renal fibrosis.

EMT IN KIDNEY FIBROSIS

Renal fibrosis represents a failed wound-healing process of renal tissue after chronic, sustained injury and is typically characterised by the remarkably consistent histopathological features of glomerulosclerosis, interstitial fibrosis and tubular atrophy implying a "final common pathway" of injury. This process is typically progressive and irreversible and the degree of interstitial fibrosis is the most important predictor of the rate of progression of chronic kidney disease (CKD) [15].

Resident renal fibroblasts were traditionally considered to be the principle mediators of fibrosis. However, convincing evidence now exists to suggest that the appearance of interstitial myofibroblasts is also critical, with myofibroblasts being more profibrotic than resident interstitial fibroblasts. Matrix producing fibroblasts and myofibroblasts are produced from differentiated epithelial cells which undergo a phenotypic conversion through EMT, a process which has increasingly been recognised as an integral part of renal fibrogenesis after injury.

The extent of which EMT contributes to renal fibrosis *in vivo* remains a matter of intense debate. Using genetically tagged proximal tubular cells in a unilateral ureteral obstruction model, Iwano et al. showed that up to 36% of all fibroblast specific protein (FSP)-1 positive fibroblasts within the interstitial space originate from renal proximal tubules, providing the most convincing evidence for EMT *in vivo* as a source of interstitial matrix-producing fibroblasts [4]. In additions, more recent studies showed, using cell lineage tracing technology, that a substantial number of interstitial fibroblasts also come from capillary endothelia through endothelial to mesenchymal transition (EndoMT) [16, 17]. EMT has been shown in various animal models of CKD including obstructive nephropathy [5, 16], diabetic nephropathy [18, 19], 5/6 nephrectomy [20, 21], experimental glomerulonephritis [22, 23] and chronic allograft nephropathy [24]. Tubular expression of mesenchymal markers such as vimentin and FSP1 have also been found in multiple progressive kidney diseases in humans including diabetic nephropathy [25], IgA nephropathy [26], chronic allograft nephropathy [24, 27], lupus nephritis [28] and pauci-immune crescentic glomerulonephritis [29]. The expression of these markers correlates well with declining renal function [25, 30]. This suggests that EMT is likely to play a role in the pathogenesis of human CKD. Hence understanding the mechanisms governing EMT is of clinical relevance for patients with CKD.

INDUCERS OF EMT AND THEIR SIGNALLING PATHWAYS IN CULTURED RENAL PROXIMAL TUBULE CELLS

The proximal tubule cells (PTC) are of mesenchymal origin and derive an epithelial phenotype during renal organogenesis through the process of mesenchymal to epithelial transition (MET). They are able to revert to a mesenchymal phenotype in response to certain physiological insults via EMT. Through this process, PTCs play an important role in the deposition of extracellular matrix. It is suggested that EMT contributes up to one-third of the total interstitial fibroblast population [4]. Several factors including high glucose, transforming growth factor (TGF)-β1, angiotensin II (AngII), and hypoxia have been shown to induce PTCs to undergo EMT through different signalling pathways. Among these, TGF-β1 is one of the most potent and well-studied factors known to be involved in all four steps of EMT, including loss of E-cadherin, de novo expression of α-SMA and actin reorganisation, disruption of the tubular basement membrane and enhanced cell migration invasion [5, 6]. TGF-β1 has been shown to activate the Notch signaling system by stimulating the expression of Notch-1 and its ligands Jagged-1 and Jagged-2 in tubular epithelial cells [7]. Using renal human primary PTCs, we have recently demonstrated that TGF-β1 repressed E-cadherin and increased α-SMA, matrix metalloproteinase (MMP)-2 and MMP9 and hence EMT and this was also mediated through Notch-1 signalling and downstream effects on snail [8].

The interrelation between AngII and TGF-β1 is well established as AngII and TGFβ share many profibrotic mediators and intracellular signalling systems. AngII also mimics many profibrotic actions of TGF-β1 and is known to play a key role in the progression of kidney disease. Many *in vitro* studies have demonstrated that AngII activates renal cells to produce profibrotic factors and extracellular matrix proteins [9, 10]. Gagliardini et al. have also demonstrated that AngII induces EMT in renal podocytes through Notch activation and increased expression of its downstream target gene hairy and enhancer of split-1(HES1) and snail upregulation [11]. Using a human podocyte cell line and human renal proximal tubular epithelial cells (HK-2 cells), Lavoz et al. has recently reported that AngII contributes to renal fibrosis independently of Notch pathway activation [12]. However, we have demonstrated that AngII induces EMT in human primary PTCs through the activation of the Notch-1 pathway in a TGFβ1 dependent and independent manner [8]. Collectively these data suggest that AngII may have Notch 1 dependent and independent effects on EMT. Our data have shown that Notch mediated effects of AngII on EMT is through snail stabilization [8]. It is not clear whether the Notch independent effects on EMT are mediated by snail.

Both high glucose and hypoxia have been shown to induce EMT through snail regulation, independent of Notch1 expression, since high glucose and hypoxia inhibited Notch1 expression in human HK2 cells. Importantly, we have demonstrated that high glucose and hypoxia induce signalling pathways that independently induce snail expression and thus EMT. Hypoxia triggers activation of hypoxia inducible factor (HIF)1α with downstream activation of lysyl oxidase-like2 (Loxl2), which is known to regulate the functional activity of snail by stabilizing its expression [13]. This effect on snail was independent of Notch1 or HES1 expression. Conversely, high glucose increased Snail through HES1, but independent of both Notch 1 and HIF1α [14]. Clearly, the above insults converge on snail upregulation, highlighting the importance of snail in mediating EMT in the kidney. A schematic diagram depicting the signalling pathways mediating snail activation in PTC induced to EMT is shown in Figure 1.

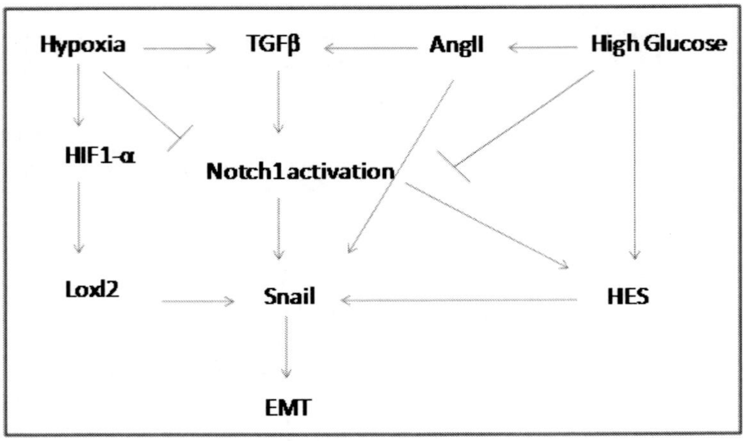

Figure 1. A simplified schematic diagram depicting the signalling pathways mediating snail activation in renal PTCs induced to undergo EMT in the presence of different insults including high glucose, hypoxia, AngII and TGFβ.

SIGNALLING PATHWAYS IN EMT- *IN VIVO* EVIDENCE

Diverse intracellular signalling pathways and mediators are involved in regulating EMT in the kidney including TGF-β/Smad, Integrin linked kinase (ILK), the canonical Wnt pathway (Wnt/β-catenin) in addition to Notch and the extracellular signal-regulated protein kinases 1 and 2 (Erk1/2) pathways.

As discussed above, TGF-β1 is known as a pivotal mediator of EMT and inducer of extracellular matrix deposition. TGFβ-induced EMT is critically relevant to the pathogenesis of kidney fibrosis. The TGFβ1/Smad signalling pathway is the most studied pathway involved in EMT. TGFβ binds to the TGFβ type I receptor which then dimerizes with the type II receptor.. This leads to downstream phosphorylation and activation of Smad2 and Smad3, which in turn heteroligomerizes with the common partner Smad4 before translocating into the nucleus. Subsequent transcription of TGFβ responsive genes, including integrin linked kinase (ILK), β1integrin, Wnt and Snail [31] then occurs. TGFβ also regulates EMT in a non Smad mediated pathway which involves p38 mitogen-activated protein kinase (MAPK) and phosphatidylinositol-3-kinase/AKt [32, 33], through which it is known to integrate with the ILK and the Wnt pathways [34].

ILK itself is an intracellular serine/threonine protein kinase which interacts with β-integrins and mediates its signalling [35]. Inhibition of ILK has been shown to attenuate renal interstitial fibrosis in a mouse model of obstructive nephropathy [35]. ILK can directly phosphorylate several physiologically important downstream effector kinases including protein kinase B (Akt) and glycogen synthase kinase 3 (GSK-3β), leading to the stabilisation of β-catenin, which in turns controls the expression of multiple genes required for EMT including T cell factor/lymphoid enhancer-binding factor-1 (LEF1), Twist, Jagged1 and Snail [31, 36].

Wnt proteins belong to a highly conserved family of secreted growth factors that play an essential role in organogenesis, tissue homeostasis and tumour formation, andhas recently been shown to promote renal fibrosis [37]. Wnt proteins signal across the plasma membrane by interacting with the Frizzled receptors and co-receptors LDL receptor–related protein-5/6. Upon binding to its receptors, Wnt proteins induce dephosphorylation of β-catenin through a series of downstream signalling events [38]. This leads to β catenin stabilization, which allows its translocation into the nucleus where it binds to LEF1 to stimulate the transcription of wnt target genes such as Twist, Jagged1 and Snail. β catenin knockdown abolishes TGFβ induced E-cadherin promoter repression, snail transcription and MMP9 expression and hence TGFβ induced EMT in renal tubular epithelial cells [39]. It is important to note that TGFβ, ILK and Wnt signals interconnect and converge at the activation of β-catenin which leads to EMT through regulation of its target genes that includes snail.

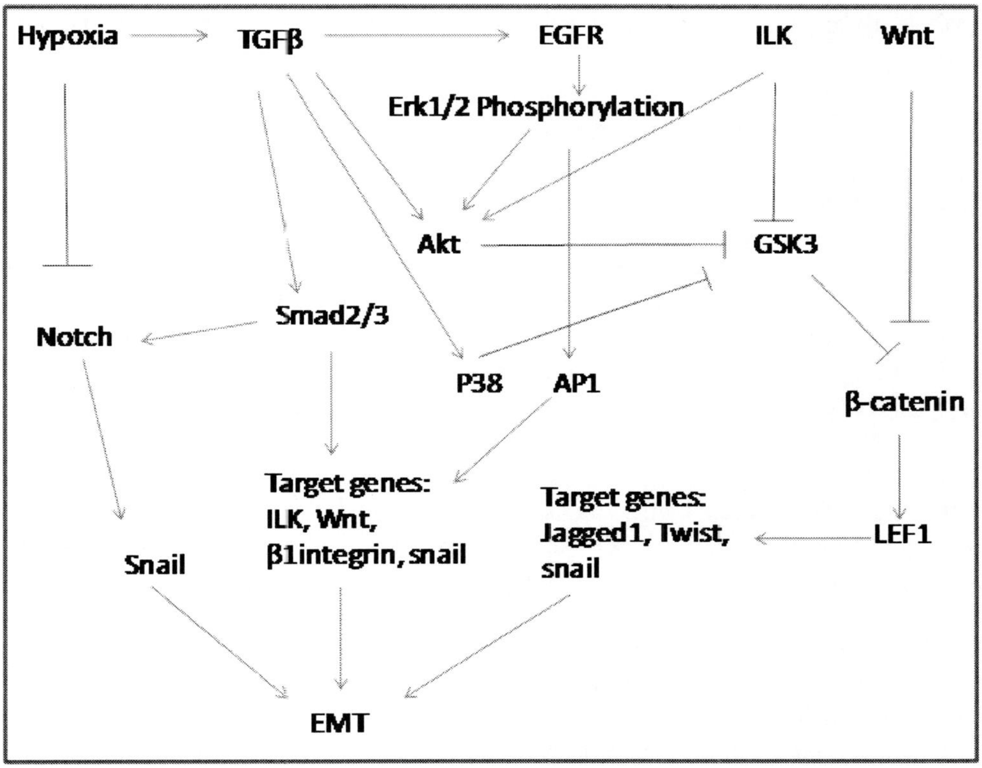

Figure 2. A simplified schematic diagram depicting the signalling pathways mediating snail activation in EMT in the kidney. EMT can be induced by a variety of stimuli and involves a diverse array of intracellular mediators which are intricately connected and integrated at different levels.

Epidermal growth factor (EGF) interacts with TGF-β to induce components of EMT in renal tubuloepithelial cells [40-42]. Activation of the EGF receptor (EGFR) tyrosine kinase can trigger EMT in tubular cells, at least in part through MAPKs signaling pathway [41, 43, 44]. Moreover, induction of EMT by oxidative or osmotic stress, and hypoxia—conditions associated with renal injury—can be accounted for in part by EGFR transactivation [45]. The transactivation of EGFR, leads to Erk1/2 phosphorylation [46]. The Erk1/2 signaling pathway plays an important role in EMT [43, 47] via the inhibition of GSK3 β and stabilization of snail [48]. Snail stability is primarily regulated by GSK3 β through the PI3K-Akt pathway [49, 50]. GSK3β phosphorylates snail and induces its nuclear exclusion, ubiquitination and subsequent degradation. Ardura et al. showed in a mouse ureteric obstruction model, that p-EGFR protein levels were augmented in the mouse kidney and this was associated with an increase of α-SMA and snail gene expression [51]. Furthermore, Erk was shown to upregulate snail gene transcription through activation of activator protein 1 (AP-1) transcription factor [52, 53] which is involved in renal fibrosis.

Thus these data demonstrate that EMT involves an intricate interplay of multiple signaling pathways, all of which seems to converge to regulate snail expression. A simplified schematic diagram depicting the signalling pathways mediating snail activation in EMT in the kidney is shown in Figure 2.

Of the many downstream target genes, snail is well characterised and highly relevant to both proteinuria and renal fibrosis, both characteristic of progressive nephropathy [54, 55].

The Role of Snail in Renal EMT

Snail is a zinc–finger transcriptional repressor, which has been characterised as a key EMT regulator. Snail is a highly unstable protein and is dually regulated by protein stability and cellular location. The Snail family includes Snail1, Snail2 (Slug) and Snail3 (Smuc). Snail regulates EMT by suppressing E-cadherin expression directly through promoter binding and repressing the kidney-specific cadherin-16, and indirectly via hepatocyte nuclear factor-1β (HNF-1β), the well characterised regulator of renal differentiation [56].

Snail induces both physiological and pathological EMT. Snail activation in the kidney promotes EMT of both tubular and collecting duct cells including fibrosis. Both epithelial cells from the renal collecting duct and the tubular epithelium are derived from snail expressing embryonic mesenchyme. Hence pathological activation of snail in the adult, which is normally repressed, can be regarded as a return to the embryonic phenotype. Increased pathological levels of Snail were shown in the fibrotic areas of human kidneys [56] and snail activation is sufficient to induce EMT, kidney injury and fibrosis in adult transgenic mice [56]. Snail is also upregulated in renal cells following UUO in mice [57]. A recent study indicated that NFkB activation leads to snail stabilization by preventing its degradation [58]. Snail inhibition also prevents podocyte injury, proteinuria and renal fibrosis in adriamycin nephropathy [59], a model of glomerular pathology. Snail is known to play an important role in modulating cell survival during EMT by inhibiting apoptosis [60-62] and regulating necrosis in response to metabolic stress [63].

Using breast cancer cell lines, Lee et al. have recently demonstrated that snail is able to inhibit mitochondrial respiration, decrease glucose consumption and lactate production and induce EMT through direct binding to the E box of the cytochrome c-oxidase (COX)VIc, COXIIa, and COXVIIc promoter regions, suggesting a new function of snail signalling in regulating mitochondrial repression and glucose metabolism through glycolytic switching [64]. The role of snail in the regulation of mitochondrial function and glucose metabolism in the kidney and whether this mediates renal induced EMT are not known but deserves investigation.

CONCLUSION

Clearly EMT is a complex process that involves many signalling pathways that can be initiated by multiple processes. Advances in imaging and cell fate technologies have confirmed the role of proximal tubular cells undergoing EMT in renal fibrosis. Intracellular signalling pathways involved in mediating EMT have been the focus of intense research interest. However, due to the complexity of the process and the interplay of multiple signalling pathways, there remains a "Holy Grail" in identifying a converging "master switch", to provide a direct target for the development of new therapeutic agents. Studies suggest the multiple signaling pathways converge on the regulation of snail expression. Snail activation induces all the features of renal fibrosis. Hence inhibiting snail activity may prove to be an effective target to prevent EMT and ultimately CKD.

REFERENCES

[1] Liu, Y., Epithelial to mesenchymal transition in renal fibrogenesis: pathologic significance, molecular mechanism, and therapeutic intervention. *J. Am. Soc. Nephrol.*, 2004. 15(1): p. 1-12.

[2] Willis, B.C., R.M. duBois, and Z. Borok, Epithelial origin of myofibroblasts during fibrosis in the lung. *Proc. Am. Thorac. Soc.*, 2006. 3(4): p. 377-82.

[3] Kalluri, R. and E.G., Neilson, Epithelial-mesenchymal transition and its implications for fibrosis. *J. Clin. Invest.*, 2003. 112(12): p. 1776-84.

[4] Iwano, M., et al., Evidence that fibroblasts derive from epithelium during tissue fibrosis. *J. Clin. Invest.*, 2002. 110(3): p. 341-50.

[5] Yang, J. and Y. Liu, Dissection of key events in tubular epithelial to myofibroblast transition and its implications in renal interstitial fibrosis. *Am. J. Pathol.*, 2001. 159(4): p. 1465-75.

[6] Fan, J.M., et al., Transforming growth factor-beta regulates tubular epithelial-myofibroblast transdifferentiation in vitro. *Kidney Int.*, 1999. 56(4): p. 1455-67.

[7] Morrissey, J., et al., Transforming growth factor-beta induces renal epithelial jagged-1 expression in fibrotic disease. *J. Am. Soc. Nephrol.*, 2002. 13(6): p. 1499-508.

[8] Saad, S., et al., Notch mediated epithelial to mesenchymal transformation is associated with increased expression of the Snail transcription factor. *Int. J. Biochem. Cell Biol.*, 2010. 42(7): p. 1115-22.

[9] Ruiz-Ortega, M., et al., Angiotensin II: a key factor in the inflammatory and fibrotic response in kidney diseases. *Nephrol. Dial Transplant*, 2006. 21(1): p. 16-20.

[10] Wolf, G., Renal injury due to renin-angiotensin-aldosterone system activation of the transforming growth factor-beta pathway. *Kidney Int.*, 2006. 70(11): p. 1914-9.

[11] Gagliardini, E., et al., Angiotensin II contributes to diabetic renal dysfunction in rodents and humans via Notch1/Snail pathway. *Am. J. Pathol.*, 2013. 183(1): p. 119-30.

[12] Lavoz, C., et al., Angiotensin II contributes to renal fibrosis independently of Notch pathway activation. *PLoS One*, 2012. 7(7): p. e40490.

[13] Peinado, H., et al., A molecular role for lysyl oxidase-like 2 enzyme in snail regulation and tumor progression. *EMBO J.*, 2005. 24(19): p. 3446-58.

[14] Sumual, S., et al., Differential regulation of Snail by hypoxia and hyperglycemia in human proximal tubule cells. *Int. J. Biochem. Cell Biol.,* 2010. 42(10): p. 1689-97.

[15] Nath, K.A., The tubulointerstitium in progressive renal disease. *Kidney Int.*, 1998. 54(3): p. 992-4.

[16] Zeisberg, E.M., et al., Fibroblasts in kidney fibrosis emerge via endothelial-to-mesenchymal transition. *J. Am. Soc. Nephrol.*, 2008. 19(12): p. 2282-7.

[17] Li, J., X. Qu, and J.F. Bertram, Endothelial-myofibroblast transition contributes to the early development of diabetic renal interstitial fibrosis in streptozotocin-induced diabetic mice. *Am. J. Pathol.*, 2009. 175(4): p. 1380-8.

[18] Burns, W.C., et al., Connective tissue growth factor plays an important role in advanced glycation end product-induced tubular epithelial-to-mesenchymal transition: implications for diabetic renal disease. *J. Am. Soc. Nephrol.*, 2006. 17(9): p. 2484-94.

[19] Holian, J., et al., Role of Kruppel-like factor 6 in transforming growth factor-beta1-induced epithelial-mesenchymal transition of proximal tubule cells. *Am. J. Physiol. Renal Physiol.*, 2008. 295(5): p. F1388-96.

[20] Lan, H.Y., Tubular epithelial-myofibroblast transdifferentiation mechanisms in proximal tubule cells. *Curr. Opin. Nephrol. Hypertens,* 2003. 12(1): p. 25-9.

[21] Liu, Y., et al., Endogenous hepatocyte growth factor ameliorates chronic renal injury by activating matrix degradation pathways. *Kidney Int.*, 2000. 58(5): p. 2028-43.

[22] Aresu, L., et al., Epithelial-mesenchymal transition (EMT) of renal tubular cells in canine glomerulonephritis. *Virchows Arch.*, 2007. 451(5): p. 937-42.

[23] Shimizu, M., et al., Role of integrin-linked kinase in epithelial-mesenchymal transition in crescent formation of experimental glomerulonephritis. *Nephrol. Dial Transplant*, 2006. 21(9): p. 2380-90.

[24] Bedi, S., A. Vidyasagar, and A. Djamali, Epithelial-to-mesenchymal transition and chronic allograft tubulointerstitial fibrosis. *Transplant Rev* (Orlando), 2008. 22(1): p. 1-5.

[25] Rastaldi, M.P., et al., Epithelial-mesenchymal transition of tubular epithelial cells in human renal biopsies. *Kidney Int.,* 2002. 62(1): p. 137-46.

[26] Nishitani, Y., et al., Fibroblast-specific protein 1 is a specific prognostic marker for renal survival in patients with IgAN. *Kidney Int.*, 2005. 68(3): p. 1078-85.

[27] Djamali, A., et al., Epithelial-to-mesenchymal transition and oxidative stress in chronic allograft nephropathy. *Am. J. Transplant*, 2005. 5(3): p. 500-9.

[28] Rossini, M., et al., Immunolocalization of fibroblast growth factor-1 (FGF-1), its receptor (FGFR-1), and fibroblast-specific protein-1 (FSP-1) in inflammatory renal disease. *Kidney Int.*, 2005. 68(6): p. 2621-8.

[29] Bariety, J., et al., Glomerular epithelial-mesenchymal transdifferentiation in pauci-immune crescentic glomerulonephritis. *Nephrol. Dial Transplant*, 2003. 18(9): p. 1777-84.

[30] Hertig, A., et al., Early epithelial phenotypic changes predict graft fibrosis. *J. Am. Soc. Nephrol.*, 2008. 19(8): p. 1584-91.

[31] Kim, M.K., et al., The differential expression of TGF-beta1, ILK and wnt signaling inducing epithelial to mesenchymal transition in human renal fibrogenesis: an immunohistochemical study. *Int. J. Clin. Exp. Pathol.*, 2013. 6(9): p. 1747-58.

[32] Bakin, A.V., et al., Phosphatidylinositol 3-kinase function is required for transforming growth factor beta-mediated epithelial to mesenchymal transition and cell migration. *J. Biol. Chem.*, 2000. 275(47): p. 36803-10.

[33] Dai, C., J. Yang, and Y. Liu, Transforming growth factor-beta1 potentiates renal tubular epithelial cell death by a mechanism independent of Smad signaling. *J. Biol. Chem.*, 2003. 278(14): p. 12537-45.

[34] Guo, X. and X.F. Wang, Signaling cross-talk between TGF-beta/BMP and other pathways. *Cell Res.*, 2009. 19(1): p. 71-88.

[35] Li, Y., et al., Inhibition of integrin-linked kinase attenuates renal interstitial fibrosis. *J. Am. Soc. Nephrol.*, 2009. 20(9): p. 1907-18.

[36] Li, Y., et al., Role for integrin-linked kinase in mediating tubular epithelial to mesenchymal transition and renal interstitial fibrogenesis. *J. Clin. Invest.*, 2003. 112(4): p. 503-16.

[37] He, W., et al., Wnt/beta-catenin signaling promotes renal interstitial fibrosis. *J. Am. Soc. Nephrol.*, 2009. 20(4): p. 765-76.

[38] MacDonald, B.T., K. Tamai, and X. He, Wnt/beta-catenin signaling: components, mechanisms, and diseases. *Dev. Cell*, 2009. 17(1): p. 9-26.

[39] Tian, X., et al., Association of beta-catenin with P-Smad3 but not LEF-1 dissociates in vitro profibrotic from anti-inflammatory effects of TGF-beta1. *J. Cell Sci.*, 2013. 126(Pt 1): p. 67-76.

[40] Strutz, F., et al., Role of basic fibroblast growth factor-2 in epithelial-mesenchymal transformation. *Kidney Int.*, 2002. 61(5): p. 1714-28.

[41] Docherty, N.G., et al., TGF-beta1-induced EMT can occur independently of its proapoptotic effects and is aided by EGF receptor activation. *Am. J. Physiol. Renal Physiol.*, 2006. 290(5): p. F1202-12.

[42] Okada, H., et al., Early role of Fsp1 in epithelial-mesenchymal transformation. *Am. J. Physiol.*, 1997. 273(4 Pt 2): p. F563-74.

[43] Grande, M., et al., Transforming growth factor-beta and epidermal growth factor synergistically stimulate epithelial to mesenchymal transition (EMT) through a MEK-dependent mechanism in primary cultured pig thyrocytes. *J. Cell Sci., 2002*. 115(Pt 22): p. 4227-36.

[44] Zhuang, S., Y. Dang, and R.G. Schnellmann, Requirement of the epidermal growth factor receptor in renal epithelial cell proliferation and migration. *Am. J. Physiol. Renal Physiol.*, 2004. 287(3): p. F365-72.

[45] Zhuang, S., et al., p38 kinase-mediated transactivation of the epidermal growth factor receptor is required for dedifferentiation of renal epithelial cells after oxidant injury. *J. Biol. Chem.*, 2005. 280(22): p. 21036-42.

[46] Syme, C.A., P.A. Friedman, and A. Bisello, Parathyroid hormone receptor trafficking contributes to the activation of extracellular signal-regulated kinases but is not required for regulation of cAMP signaling. *J. Biol. Chem.*, 2005. 280(12): p. 11281-8.

[47] Rhyu, D.Y., et al., Role of reactive oxygen species in TGF-beta1-induced mitogen-activated protein kinase activation and epithelial-mesenchymal transition in renal tubular epithelial cells. *J. Am. Soc. Nephrol.*, 2005. 16(3): p. 667-75.

[48] Cho, K.B., et al., Overexpression of c-myc induces epithelial mesenchymal transition in mammary epithelial cells. *Cancer Lett.*, 2010. 293(2): p. 230-9.

[49] Song, L.B., et al., The polycomb group protein Bmi-1 represses the tumor suppressor PTEN and induces epithelial-mesenchymal transition in human nasopharyngeal epithelial cells. *J. Clin. Invest.,* 2009. 119(12): p. 3626-36.

[50] Zhou, B.P., et al., Dual regulation of Snail by GSK-3beta-mediated phosphorylation in control of epithelial-mesenchymal transition. *Nat. Cell Biol.*, 2004. 6(10): p. 931-40.

[51] Ardura, J.A., et al., Parathyroid hormone-related protein promotes epithelial-mesenchymal transition. *J. Am. Soc. Nephrol.,* 2010. 21(2): p. 237-48.

[52] Hudson, L.G., et al., Ultraviolet radiation stimulates expression of Snail family transcription factors in keratinocytes. *Mol. Carcinog*, 2007. 46(4): p. 257-68.

[53] Li, Y., et al., UV irradiation induces Snail expression by AP-1 dependent mechanism in human skin keratinocytes. *J. Dermatol. Sci.*, 2010. 60(2): p. 105-13.

[54] Dai, C., et al., Wnt/beta-catenin signaling promotes podocyte dysfunction and albuminuria. *J. Am. Soc. Nephrol.*, 2009. 20(9): p. 1997-2008.

[55] Matsui, I., et al., Snail, a transcriptional regulator, represses nephrin expression in glomerular epithelial cells of nephrotic rats. *Lab. Invest,* 2007. 87(3): p. 273-83.

[56] Boutet, A., et al., Snail activation disrupts tissue homeostasis and induces fibrosis in the adult kidney. *EMBO J.*, 2006. 25(23): p. 5603-13.

[57] Sato, M., et al., Targeted disruption of TGF-beta1/Smad3 signaling protects against renal tubulointerstitial fibrosis induced by unilateral ureteral obstruction. *J. Clin. Invest.*, 2003. 112(10): p. 1486-94.

[58] Wu, Y., et al., Stabilization of snail by NF-kappaB is required for inflammation-induced cell migration and invasion. *Cancer Cell,* 2009. 15(5): p. 416-28.

[59] He, W., et al., Blockade of Wnt/beta-catenin signaling by paricalcitol ameliorates proteinuria and kidney injury. *J. Am. Soc. Nephrol.*, 2011. 22(1): p. 90-103.

[60] Vega, S., et al., Snail blocks the cell cycle and confers resistance to cell death. *Genes Dev*, 2004. 18(10): p. 1131-43.

[61] Escriva, M., et al., Repression of PTEN phosphatase by Snail1 transcriptional factor during gamma radiation-induced apoptosis. *Mol. Cell Biol.*, 2008. 28(5): p. 1528-40.

[62] Kurrey, N.K., et al., Snail and slug mediate radioresistance and chemoresistance by antagonizing p53-mediated apoptosis and acquiring a stem-like phenotype in ovarian cancer cells. *Stem Cells*, 2009. 27(9): p. 2059-68.

[63] Kim, C.H., et al., Implication of snail in metabolic stress-induced necrosis. *PLoS One*, 2011. 6(3): p. e18000.

[64] Lee, S.Y., et al., Wnt/Snail signaling regulates cytochrome C oxidase and glucose metabolism. *Cancer Res.*, 2012. 72(14): p. 3607-17.

INDEX

C

D

E

F

I

J

K

L

T

U

V

W

X

Y

Z